te 3. CIRROCUMULUS

Plate 4. ALTOSTRATUS

te 7. STRATOCUMULUS

Plate 8. NIMBOSTRATUS

te 11. CUMULONIMBUS

Plate 12. MAMMATUS

Weather
for
Outdoorsmen

Also by Walter F. Dabberdt
THE WHOLE AIR WEATHER GUIDE

Walter F. Dabberdt

WEATHER FOR OUTDOORSMEN

A Complete Guide to Understanding and Predicting Weather in Mountains and Valleys, on the Water, and in the Woods

CHARLES SCRIBNER'S SONS • *New York*

To Meredith, Jennifer, and Geoffrey,
for their help and understanding
and to SRI International, for your
support and encouragement

Copyright © 1981 Walter F. Dabberdt

Library of Congress Cataloging in Publication Data

Dabberdt, Walter F.
Weather for outdoorsmen.

Bibliography: p.
Includes index.
1. Weather.   I. Title.
QC861.2.D3   551.6'3   81–228
ISBN 0-684-16865-0   AACR2

1 3 5 7 9 11 13 15 17 19 Y/C 20 18 16 14 12 10 8 6 4 2

Printed in the United States of America

Charts and drawings by John Zermani

Contents

ENDPAPER ILLUSTRATIONS

Plate 1. Cirrus. *High and thin, fair weather; low and wispy, a weather change is possible.* Courtesy of the National Oceanic and Atmospheric Administration

Plate 2. Cirrostratus. *Rain possible tomorrow when cirrostratus follow cirrus and precede altostratus.* Courtesy of Fritz Krügler

Plate 3. Cirrocumulus. *A transient cloud usually accompanied by fair weather.* Courtesy of Fritz Krügler

Plate 4. Altostratus. *Rain likely in 6–12 hours when altostratus follow cirrus and cirrostratus.* Courtesy of the National Oceanic and Atmospheric Administration

Plate 5. Altocumulus. *Usually a sign of fair weather, unless immediately followed by towering cumulus.* Courtesy of the National Oceanic and Atmospheric Administration

Plate 6. Stratus. *Expect only drizzle.* Courtesy of the National Oceanic and Atmospheric Administration

Plate 7. Stratocumulus. *Often precede a cool, clear evening.* Courtesy of Fritz Krügler

Plate 8. Nimbostratus. *Prolonged, steady precipitation is likely.* Courtesy of Fritz Krügler

Plate 9. Cumulus. *A sign of fair weather for the day.* Courtesy of the National Oceanic and Atmospheric Administration

Plate 10. Cumulus Congestus. *Expect rain showers and, if seen by midmorning, an afternoon thunderhead.* Courtesy of the National Oceanic and Atmospheric Administration

Plate 11. Cumulonimbus. *Heavy rain showers and gusty winds; expect thunder and lightning with a chance of hail.* Courtesy of the National Oceanic and Atmospheric Administration

Plate 12. Mammatus. *Not a cloud type, but rather a severe-weather feature of several clouds.* Courtesy of the National Oceanic and Atmospheric Administration

Preface

"Why a weather book for outdoorsmen?" an amateur sailor friend asked me recently. "After all, most of us going into the outdoors already know just about all we need to about the weather. And what could you possibly tell us about the things we really need to know—such as, when will we get forecasts that we can believe and trust?" His comments and questions are probably typical of most boaters, hikers, skiers, fishermen, and other outdoor enthusiasts. Unfortunately, I didn't need to think very long before replying. I say unfortunate because two days earlier two 30-foot-plus sailboats with auxiliary power had set out from San Francisco for a holiday sail some 100 miles down the coast to Monterey. They never made it. Forced to turn back by the heavy winds and high seas of an early winter storm, they broached and capsized, killing four people. The forecast was not at fault; perfectly adequate weather information and forecasts were readily available.

The information in this book could have been the missing ingredient that stormy day as it could be in countless other weather tragedies. I hope that when you venture into the elements, you will do so with a genuine respect for weather and that you will have the know-how to anticipate the worst and take timely and effective actions to avoid or minimize dangers.

While such information is the book's primary purpose, there is a second, equally important objective. Weather can be as tranquil as it is tormenting, as beautiful as it is destructive, as stimulating as it is frustrating. If you aren't aware of the fascinating characteristics of the air around us, perhaps these pages will inspire you in that direction.

This book is intended as a practical, how-to manual, and there are several logical ways you can use it. You can even skip the first five chapters if what interests you most is in the sixth. The material is organized to help you locate and digest the weather information you need, when you need it. The introduction provides a glimpse of what weather is and the ways it can affect outdoorsmen (including women, of course). It answers questions such as: Why study weather? What is climate? How do human beings affect climate? Part One provides a basic and comprehensive discussion of weather and climate. Part Two covers the specifics of weather. Weather conditions particular to each of three basic outdoor environments are discussed: oceans and lakes; mountains and valleys; and forests. These three chapters discuss weather conditions that may affect you significantly if you are in an unsheltered environment. The emphasis is not so much on "turning on the TV and running for the storm cellar" as it is on knowing what to expect, how to foretell its happening, and how to minimize or avoid the consequences.

Four appendices provide practical information for the weather-conscious person. The Glossary gives plain-language definitions for more than one hundred weather terms. Weather and Your Health and Comfort deals with the effects of heat, cold, wind, and humidity. Instant Weather Information tells where and how to get current forecasts and weather data, and A Backpack Weather Station describes the simple and affordable gear that will help you collect useful weather information in the outdoors. But before "diving in" and becoming an ardent amateur weathercaster, pause first to consider this humbling piece of anonymous prose:

> *Many critics, few defenders,*
> *Weathermen (even amateurs) have but two regrets:*
> *When they hit, no one remembers;*
> *When they miss, no one forgets.*

WALTER F. DABBERDT
Los Altos, California

Introduction

What time is it now as you read this? If it's evening, think for a moment about today's weather. If it's earlier in the day, recall yesterday's. Next, remember the previous day. As you do so, try to answer the following simple questions that will test your weather awareness:

- What types of clouds were there this morning? This afternoon? How did they change: Did they become lower, thicker, or more widespread?
- When did the wind pick up? Or was it already blowing around sunrise? What direction was it? Did the direction change throughout the day or night?
- Was there a halo around the moon last night? Did high-flying jets leave condensation trails behind?
- What was the visibility at midday? Five, ten, twenty-five or more miles? Less?
- What was the highest temperature during the day? When did it occur?
- How humid was the air? Less than 40 percent relative humidity? More than 75 percent?
- Did it rain or snow? When did it start and end? Was there fog?

If you had difficulty answering these questions, you need to become more weather-conscious. Reading these pages in the shelter of your home or office, you have little to fear from being unaware of the weather and not anticipating its changes. You might forget an umbrella or sweater and get wet or chilled, but probably nothing more severe will happen to you. But if you're unprotected on the water, in the woods, or on a mountain face, the penalty may be much harsher. Weather-sense is not something most people can turn on and off at will. If you don't have it every day, you probably won't have it when you really need it: when your life or the lives of those around you may depend on it.

What is weather? Very simply, it is the condition or state of the atmosphere at any instant of time. It's the evening land breeze that blows offshore in summer. It's valley fog—or hurricanes and tornadoes. Weather can bring peace and serenity one day; pain and destruction the next.

Weather differs from climate. Weather describes atmospheric conditions for a minute, an hour, or a day, whereas climate is the average of weather conditions for a long period of time—at least thirty years. Saying "the climate of the seventies was different from the climate of the fifties" is incorrect. Ten years is too short a period to properly define climate. The more proper reference would have been to the "average weather" of the fifties or seventies. If weather is a snapshot, then climate is the time exposure.

Our atmosphere is a complex and often poorly understood weather machine. Fortunately, we know enough about how it works (if not always precisely why) to predict reliably the more significant weather features one or two days in advance. The computers at the National Weather Service have a batting average of about 83 percent for twenty-four-hour forecasts. However, it is usually the one miss in six that we remember most! But we *can* avoid or minimize the impact of these misses by knowing how weather changes; we may then be able to recognize when a forecast is failing and what changes to expect.

We can look at this concept another way. Often the forecasters at the National Weather Service will believe that any one of two or three weather alternatives is probable over the next twenty-four hours. As an example, they may consider that an advancing warm front will either (1) continue along its present path, (2) intensify as it slows down, or (3) change course and bypass the area altogether. The forecasters may rate the first alternative as the most likely and give it a 50 percent chance or probability to occur—meaning that with current weather conditions there are five chances in ten the forecasted weather will occur (it will rain tomorrow).* Therefore, knowing and understanding the forecast and its corresponding probability permits the weather-conscious person to judge when a forecast is going awry, and, more importantly, to assess which of the other weather alternatives is then most probable. So if a forecast calls for clear skies and light northerly winds and you observe southerly winds with altostratus clouds preceded by cirrostratus and cirrus, you'll recognize that the frontal system has probably moved farther north than predicted and that rain is likely within the next six to twelve hours.

There is more to understanding and predicting the weather than tracking the movement of large-scale (synoptic) storm systems and air masses. Weather operates on a variety of scales from large to small. Some of the time the large-scale weather patterns dominate and control local weather conditions; at other times it is the small-scale (or microscale) effects that are important. As an example, coastal navigation may be hampered in the morning by ground fog that has formed over land and drifted just offshore. However, the same high-pressure system that caused the fog to form over land in the first place can also produce clear skies and unrestricted visibility farther out over open water.

In the chapters that follow, one central aim will be pursued: the broadening and honing of your weather-consciousness and the

---

*There is yet another way to think of probability. Using the "50 percent example," we can say that in the past, the same set of currently observed weather conditions produced the forecasted conditions 50 percent of the time.

development of an awareness of the impact weather has on you and your outdoor activities. You won't acquire weather-consciousness all at once, either through the chapters that follow or in the out-of-doors. Because weather has a "memory," you must understand today's weather in relation to yesterday's to predict tomorrow's. You must also look for telltale signs in the winds and the clouds, and then piece together for yourself the parts of the weather puzzle. But the effort will offer two very tangible rewards: By better understanding the published weather forecast, you'll be able to do a better job of *planning* for tomorrow's activities. And once under way, your increased understanding of current weather conditions and their significance will lead to a safer and more enjoyable adventure—better *execution.* Planning and execution—these are the two ways your improved weather knowledge will help. You'll know that cumulus congestus clouds in midmorning indicate a good possibility for afternoon thunderstorms, while a field of cumulus humulus is a sure sign of continued fair weather and moderate winds. Seeing the latter can mean planning on a good-weather day; the former, planning to execute a hasty run for shelter.

We in the twentieth century are far from being the first to try to understand the weather. The earliest known weather forecasters or meteorologists (of sorts) are thought to have been the ancient Egyptians, who may be considered embryonic meteorologists because of their use of rainmakers around 3500 B.C. Unfortunately, we don't know their track record. A more serious study of meteorology began in Greece around 550 B.C. with the works and theories of Thales of Miletus and flourished for five centuries until halted by the conquering Romans. Two protégés of Thales, Anaximander and Anaximenes, gave the first, if faulty, explanation of thunder and lightning: Thunder was the sound of air clashing against the sides of clouds, while air rushing through the cloud "kindled the flame of lightning." On a more promising note, Anaximander was the first to properly define wind as "a flowing of air." A century later the last follower of Thales, Anaxagoras, put forth two revolutionary weather theories. In one he proposed that clouds contained fire, or "aether," which originated in the upper atmosphere. Lightning then followed as this fire flashed through the clouds, where it was subsequently quenched and produced thunder. His other theory is less romantic, though scientifically more accurate: Hail, he said, was formed by water freezing as it was pushed upward within the clouds. (Interestingly, this second theory apparently contradicts the first, which assumed there was fire both above and within the clouds.)

Over the next two centuries the Greeks pursued their meteorological studies. Perhaps the most profound theory of this period was proposed by Democritus (460–357 B.C.): that storm systems actually move from place to place. However, this theory was not generally believed until it was rediscovered more than 2,000 years later by Benjamin Franklin.

A major historical event occurred around 340 B.C. when Aristotle (384–322 B.C.) wrote the first comprehensive weather treatise, titled *De Meteorologica.** This four-volume work summarized previous weather theories and principles and added some new ones. The foundation of Aristotle's theories was his belief that the terrestrial universe (planet Earth) consisted of four elements. These were thought to be oriented in four concentric spheres or rings, with earth at the core and water, air, and fire successively above. Aristotle did, however, recognize exceptions to this orientation; for example, mountains (earth) can be above water, fire may be within clouds (air), and so forth. Unlike some of his predecessors whom he took to task in his book, Aristotle used the four-element theory in an attempt to describe observed phenomena, rather than using

---

*An English translation by H. D. P. Lee was published in 1952 by Harvard University Press.

observations to deduce physical theories. The latter is called deductive reasoning, and is a major and necessary part of understanding weather, particularly when you are isolated in the outdoors. There, observations may be your *only* key to understanding local weather conditions.

Errors and omissions aside, Aristotle's *De Meteorologica* stood as the only comprehensive compendium of meteorology for nearly two millennia—until the seventeenth century A.D. For the most part, new concepts in meteorology were not developed during this period; rather, things were explained by means of the terms and theories of Aristotle.

One of Aristotle's pupils, Theophrastus of Eresos (*c.* 372–287 B.C.), wrote two interesting books of proverbs for forecasting bad weather; these followed by a few years a similar though less extensive work by the pre-Aristotelian philosopher Eudoxus of Gnidos from Babylon. Theophrastus is acknowledged as the author of everyone's first weather rhyme:

> Red sky at morning, sailor's warning.
> Red sky at night, sailor's delight.

This particularly reliable proverb is based on the fact that storms generally move from west to east. The red evening sun is caused by the atmospheric dust that accompanies dry weather. Since the sun sets in the west, the red sunset indicates fair weather in that direction. The opposite situation in the morning indicates that poor weather is in the area, and its evening winds have kept the air well mixed with the light-scattering particles that produce the red color. This particular proverb is thought to have been widely known in coastal areas after the time of Theophrastus. The best-known figure to use the proverb publicly was Jesus Christ, as described in Matthew 16:2–3. Theophrastus also included proverbs more oriented toward the farmer, such as the less familiar:

It is a sign of storm or rain
When the ox licks his forehoof.

As one usually has no ox while backpacking or hunting, I'll assume this rhyme to be of less interest and forgo discussion of its merit.

In the 1600s, weather instruments were first devised, and meteorologists could now objectively *measure* temperature, pressure, humidity, and wind speed. With these tools, meteorology took its first step from art to science. Ben Franklin was its American pioneer, discovering the nature of lightning around 1750, and later publishing weather outlooks in his *Poor Richard's Almanack*. Weather maps (called synoptic charts) were first drawn in the early nineteenth century and, with them, meteorologists had their first glimpse of large-scale weather systems. Daily forecasts were first prepared on September 22, 1869, by Cleveland Abbe for the Cincinnati Chamber of Commerce. Six months later, the federal government organized a national weather agency—in part prompted by the loss of and damage to 1,164 ships the preceding winter on the Great Lakes alone. This agency has since grown to today's National Weather Service, which maintains a network of almost 4,000 land and marine weather-observing stations. These weather observations, together with those from satellites, balloons, aircraft, and radar, go into producing some form of weather advisory every sixteen seconds in the United States.

PART ONE

Understanding Weather and Climate

Chapter One

The Elements of Weather

Weather is a dichotomy; on the one hand, it can be a difficult concept to understand and comprehend, while on the other, it is easily recognized and appreciated. We all recognize what weather is and the effects it can have on our lives and our environment. We feel its warmth in summer and try to avoid its icy bite in winter. We don't really see the wind but feel it on our faces and watch the trees sway in it. We sense that certain types of weather help us feel more alert and responsive; other types seem to make us apprehensive and tense. Other effects of weather are better known to us: the wind and snow of blizzards, flash flooding, summer heat waves and brownouts, hurricanes and tornadoes, thunderstorms, lightning, hailstorms, and so on.

To better understand how bad (and good) weather comes about, and how we can even predict it ourselves, it is necessary first to understand the elements that comprise weather. In this regard, the attempt to understand weather is like the efforts of the three blind men of the old Indian parable: Each tried to describe an elephant by touching a different part of its body. One touched its trunk, the second its ears, and the third its feet. Each had an accurate understanding of the one part he examined, but it wasn't until the three got together and exchanged their knowledge that a total (if not totally accurate) picture of the elephant emerged. In the same way, the weather-conscious naturalist needs first to understand each of several elements (temperature, pressure, humidity, and so forth) before piecing them together to make a complete model of the atmosphere.

Temperature

Perhaps the most important weather element is temperature. It is simply the hotness or coldness of an object as measured by a thermometer. But temperature has yet a more fundamental importance in that it is a direct indicator of the amount of heat energy contained in that object. So when we say the temperature of water is 33°F, we are also saying implicitly that the heat energy of the water is sufficient to keep it liquid, whereas if we removed enough heat, the water would freeze and become ice as its temperature dropped below 32°F.

Temperature and heat go hand in hand. Heat content can be expressed as *calories, BTU's, joules,* or a host of other techno-units, whereas temperature can be given as *Fahrenheit, Celsius* (or *Centigrade*), *Kelvin, Absolute,* and a few others no longer in use. In the United States and Canada the Fahrenheit scale has been the most common, although it is gradually being replaced by the Celsius scale, which is dominant in most of the rest of the world. The United States started converting in 1975 to the Celsius scale for use in weathercasting. The Celsius and Centigrade scales are identical, except that the term "Centigrade" was officially abandoned in favor of "Celsius" by the 1948 (Ninth) General Conference on Weights and Measures.

In the Fahrenheit scale water freezes at 32°F and boils at 212°F. But what is the significance of 0°F and 100°F? A temperature of 0°F is the lowest at which salt will prevent water from freezing. The 100°F mark has no apparent basis, although it is close to body temperature. Gabriel Daniel Fahrenheit (1686–1736), the German physicist who first used mercury in thermometers and developed the temperature scale bearing his name in 1710, originally determined that the normal temperature of the human body was 96°F; later studies proved it to be 98.6°F.

The Celsius temperature scale has two readily remembered points: 0°C is the temperature at which pure water freezes, and 100°C is the boiling point of water. This scale was developed in 1742 by the Swedish astronomer Anders Celsius (1701–1744), who originally set the boiling and freezing points in reverse position from what they are today (that is, he took the boiling point as 0°C and the freezing point as 100°C). On the Celsius scale, human body temperature is exactly 37°C. Fahrenheit and Celsius temperatures are equal at −40°.

Temperature is a good indicator of body comfort. While the ideal air temperature may be around 80°F (26.7°C), it is rarely that. Air temperatures are usually measured about six feet above the ground in a shelter that shades the thermometer from the sun. In the United States, air temperatures have ranged from a high of 134°F (56.7°C) on July 10, 1913, at Death Valley, California, to a low of –76°F (–60°C) during January 1886 at Tanana, Alaska.

Temperatures ultimately are controlled by the amount of energy received from the sun. Long summer days with "high" sun are obviously a lot warmer than short winter days with the sun low in the sky. Clouds modify temperatures by reducing the portion of the sun's energy that reaches the ground; clouds do this by reflecting a large fraction back to outer space and absorbing some of the energy themselves. Apart from such effects as these, temperatures vary widely depending on other important factors: the physical makeup of the ground, slope, latitude, altitude, and something meteorologists call advection.

For the time being, we will ignore the effects that trees have on air temperatures near the ground and focus instead on the nature and makeup of the ground surface. To prove to ourselves the importance of ground effects, we only need to recall the temperature difference that exists on calm days between ocean and beach, or between asphalt and grass. Differences occur because the thermal properties of these various substances are all very different. Two properties in particular have a major effect on temperature. One of them is called *heat capacity* or *specific heat,* the other is *thermal conductivity.* Heat capacity is the amount of heat energy (measured in calories) that must be supplied to raise the temperature of the substance a specified amount (say, 1°C). Thermal conductivity is the ability of the substance to conduct heat (or, put another way, its ability to transport the heat away from the location where the energy is applied).

Now, if all this sounds very technical and esoteric, let's explore the practical aspects. Water has a large heat capacity; in other words, a lot of energy is needed to raise the temperature of water. The thermal conductivity of water depends on whether it is calm or in motion. Still water does not transmit heat very readily. Silver conducts heat almost 1,000 times faster than still water, which in turn is some five times faster than dry wood and twice as fast as dry sand. But if water is in motion, its conductivity increases accordingly. Calm air, in contrast, is a very poor conductor of heat, although it requires only a very small amount of heat to raise its temperature. Accordingly, air is an effective insulator. Body warmth is not readily conducted through the air layer in the lining of a down jacket or sleeping bag nor through the hay-type mulch used to protect sensitive plants from early-winter frosts. Temperature variations reflect the combined effects of heat capacity and thermal conductivity. The surface of a poor conductor with a low heat capacity will heat up quickly when exposed to the sun and will cool off just as rapidly when the heat source is removed. On the other hand, good conductors with large heat capacity heat up and cool off little.

Table 2 (page 8) shows the relative thermal properties of some common materials and soils along with the responsiveness of each to the addition or removal of heat. Table 3 presents temperatures that were actually observed for each of six different ground covers during the months of June and January. The asphalt surface has the warmest daytime temperatures and the largest daytime-nighttime

TABLE 1. FAHRENHEIT–CELSIUS TEMPERATURE CONVERSION CHART

Temperatures inside the table are Celsius; those in the column and row headings are Fahrenheit. Example One: A temperature of −15°F is tabulated by the intersection of the horizontal row marked ″−10″ and the vertical column marked ″−5″ and is equivalent to −26.1°C (Celsius). Example Two: +63°F = 17.2°C, as given by the value at the intersection of the +60 row and the +3 column.

Sub-Zero Temperatures (°F)	0	−1	−2	−3	−4	−5	−6	−7	−8	−9
−40	−40.0	−40.6	−41.1	−41.7	−42.2	−42.8	−43.3	−43.9	−44.4	−45.0
−30	−34.4	−35.0	−35.6	−36.1	−36.7	−37.2	−37.8	−38.3	−38.9	−39.4
−20	−28.9	−29.4	−30.0	−30.6	−31.1	−31.7	−32.2	−32.8	−33.3	−33.9
−10	−23.3	−23.9	−24.4	−25.0	−25.6	−26.1	−26.7	−27.2	−27.8	−28.3
0	−17.8	−18.3	−18.9	−19.4	−20.0	−20.6	−21.1	−21.7	−22.2	−22.8

Above-Zero Temperatures (°F)	0	+1	+2	+3	+4	+5	+6	+7	+8	+9
0	−17.8	−17.2	−16.7	−16.1	−15.6	−15.0	−14.4	−13.9	−13.3	−12.8
+10	−12.2	−11.7	−11.1	−10.6	−10.0	−9.4	−8.9	−8.3	−7.8	−7.2
+20	−6.7	−6.1	−5.6	−5.0	−4.4	−3.9	−3.3	−2.8	−2.2	−1.7
+30	−1.1	−0.6	0.0	0.6	1.1	1.7	2.2	2.8	3.3	3.9
+40	4.4	5.0	5.6	6.1	6.7	7.2	7.8	8.3	8.9	9.4
+50	10.0	10.6	11.1	11.7	12.2	12.8	13.3	13.9	14.4	15.0
+60	15.6	16.1	16.7	17.2	17.8	18.3	18.9	19.4	20.0	20.6
+70	21.1	21.7	22.2	22.8	23.3	23.9	24.4	25.0	25.6	26.1
+80	26.7	27.2	27.8	28.3	28.9	29.4	30.0	30.6	31.1	31.7
+90	32.2	32.8	33.3	33.9	34.4	35.0	35.6	36.1	36.7	37.2
+100	37.8	38.3	38.9	39.4	40.0	40.6	41.1	41.7	42.2	42.8

TABLE 2. THERMAL CHARACTERISTICS OF SOME NATURAL AND MAN-MADE MATERIALS

Type of Soil or Material	Physical Characteristics		Magnitude of the Temperature Response to an Input of Heat
	Heat Capacity	Thermal Conductivity	
Silver	M	VL+	VS
Iron	L	VL	VS
Concrete	M	L	M
Rock	M	L	M
Ice	M	M	M
Wet sand	M	M	M
Old snow	M	M	M
Still water	VL	M	M
Dry sand	S	S	L
New snow	S	S	L
Dry wood	S	S	L
Dry moors	S	S	L
Still air	VS−	VS	VL

VS = Very small
S = Small
M = Moderate
L = Large
VL = Very large

TABLE 3. SOME EXAMPLES OF THE VARIATION IN TEMPERATURE* OF DIFFERENT TYPES OF SURFACES DURING ONE DAY FOR SUMMER AND WINTER

Type of Ground or Surface	Summer (June) Temperatures (°F)			Winter (January) Temperatures (°F)		
	High	Low	Daily Variation	High	Low	Daily Variation
Asphalt	109°	50°	59°	47°	34°	13°
Sandy soil	95°	48°	47°	45°	36°	9°
Bare earth	96°	51°	45°	45°	35°	10°
Gravelly soil	88°	50°	38°	45°	35°	10°
Beneath grass	85°	56°	29°	43°	37°	6°
Loam	76°	56°	20°	44°	35°	9°
Air, 4 feet above ground	71°	46°	25°	47°	35°	12°

SOURCE: N. K. Johnson and E. L. Davies, "Some measurements of temperature near the surface in various kinds of soils," *Quarterly Journal of the Royal Meteorological Society*, vol. 53, 1927, pp. 45–59.

*All ground temperatures were recorded ³/₈ in. below the surface.

variation in both months. The grass and loam surfaces are the poorest conductors of the lot and have the warmest evening temperatures, but also the coolest daytime temperatures. These examples indicate that nighttime temperatures at an open campground can vary by up to 15°F at ground level between sites with different ground covers (for example, soil versus grass). That's something to keep in mind when looking for a location for your sleeping bag.

The nature of the ground surface not only strongly influences the temperature highs and lows, but also controls the time that these extremes occur. Right at the ground surface, temperatures reach their peak about two hours later than the time when the sun is strongest (called solar noon). Air temperatures a few feet above ground typically reach a maximum about two to three hours later than the time of peak ground surface temperatures. The temperature variation below the ground surface is similar to that at the surface, except that the delay increases with depth, while the magnitude of the variation decreases rapidly with depth below the surface. These properties can be extremely beneficial to someone seeking relief from the extreme cold of a winter blizzard. More on this aspect in Part Two.

Temperature varies not only with surface conditions, time of day, season, and latitude, but also with altitude. On the average, temperatures decrease by 3.5°F (1.7°C) for every 1,000-foot increase in elevation up to about 30,000 feet. The actual rate at which temperature decreases with height is called the *temperature lapse rate*. But the lapse rate you encounter while climbing, flying, or ballooning will usually be far different from the average value. With overcast skies, the temperature is more likely to drop about 5.5°F (3°C) for every 1,000 feet of elevation; on sunny summer days, the drop will be even greater. Clear, calm nights lead to an opposite situation in which the temperature actually increases with height. This condition is called a *temperature inversion* and is one of the principal contributors to the smoggy conditions of cities such as Los Angeles, Denver, San Francisco, and Fairbanks.

(To be totally accurate, there are many causes of temperature inversions apart from the dramatic cooling off of the ground and adjacent air layers under clear, calm nighttime skies. The semipermanent inversions indigenous to the West Coast are created by the heating of air at altitudes of 1,500 feet and above.)

One of the more dramatic causes of temperature change is *advection.* Although it is not a common term, instances and impacts of advection are well known to us all. Advection refers to the importation of air from another region where the air has characteristics unlike our own. Air that is transported to the northeastern United States in summer from the Gulf of Mexico brings with it high temperatures (and equally high humidities). "The Hawk, Mr. Wind," is another case of advection common to Chicago, where cold air sweeps in either from Canada or off Lake Michigan. Advection and its effect on temperature will be examined further in the discussion of air masses in Chapter 4.

Pressure

One of the most widely used terms in weather, pressure is perhaps the least understood and the most widely misrepresented—at least on the airwaves. Let's start with a basic definition: Pressure is the force that the weight of the overlying atmosphere exerts on the ground, people, or buildings. Theoretically speaking, our atmosphere continues until there are no more molecules of oxygen, nitrogen, helium, carbon dioxide, and so on. Thought of this way, the atmosphere continues virtually to infinity, although in reality it merges beyond a height of 500 miles with the lonely hydrogen and helium molecules that make up what is sometimes called the *interplanetary gas.* The atmosphere is most dense at or close to ground level and thins out at an increasing rate with altitude. Fifty percent of the atmosphere is below 18,000 feet. Transcontinental jet airliners flying at 36,000 feet have 75 percent of the atmosphere below them (said a different way, the air pressure outside the plane is about 25 percent of what it is at sea level). Above 54,000 feet (near the level of trans-Pacific 747-SP jet planes) only 10 percent of the atmosphere remains, while only 1 percent is found above 20 miles (105,000 feet). Astronaut wings are not awarded until 99.999 percent of earth's air lies beneath one at a height of 50 miles!

The atmosphere exerts pressure because air is heavy. A cubic yard of the stuff weighs about 2 pounds at sea level, while the weight of all of the earth's atmosphere has been estimated to be 5,600 trillion tons. This is the same weight that would result if the earth were totally covered by a shell of water 34 feet thick. Before exploring some of the consequences of pressure (and its variation from place to place or day to day), let's take a quick look at how pressure is reported and measured.

"The six o'clock sea level pressure is 29.92," reports our TV weather forecaster. We have a hunch that's sort of normal, and although we're not sure how to use this piece of information, we are somewhat reassured for having it. Let's correct this bit of vaguery and make the pressure reading a useful bit of information. The "29.92" refers to the height in inches of a column of mercury in a mercurial barometer. This pressure instrument measures the pressure exerted on it by the overlying column of air (which has its bottom at the height of the barometer and its top in interplanetary space). At the moment of the report the overlying atmosphere had a weight equal to the weight of 29.92 inches of mercury. If mercury weren't available, we could theoretically use water in a barometer—except it would give us a column almost 34 feet tall. Some other pressure units and the values they would have for this example include millibars (1,013.2), millimeters of mercury (760), and pounds per square inch (14.7).

Pressure readings are always reported as if the barometer were located at sea level, even if the measurement is made in mile-high Denver. This is done by applying a correction factor that accounts for the height of the instrument. Within a few thousand feet of sea level, pressure decreases about one inch of mercury (the chemical symbol is Hg) for every 1,000-foot increase in altitude. Referring to the example, the 29.92 pressure reading is understood to be the sea-level equivalent.

The pressure reading of and by itself tells us very little about the weather and what changes to expect. The pressure information we really need to know is:

- The *pressure tendency,* or the change in pressure that has occurred over the past six to twelve hours (that is, whether the pressure has increased or dropped, and how much of a change took place)
- The *pressure gradient,* or the difference in pressure that exists between our location and adjacent areas

As we will see later, an increase in pressure (sometimes termed a "rising barometer") indicates that fairer, drier weather is on the way. The faster the pressure changes, the more probable is the fair weather and the sooner it can be expected. In the opposite way, decreasing pressure, or a falling barometer, is one warning of an approaching storm. The more intense storms—those with higher winds and larger amounts of precipitation—are normally accompanied by a very large decrease in pressure. A pressure drop of about .3 inch in a six-hour period should be considered large and a reliable indicator of impending bad weather.

In a similar way, the pressure gradient is a useful indicator of weather conditions. On a weather map, lines connecting points of equal pressure are called *isobars.* The separation of isobars determines the pressure gradient. If the isobars are close together, it means that the pressure at neighboring locations is very different and the pressure gradient is said to be large or "strong." A small or "weak" gradient means that pressures are more or less the same everywhere. Sometimes the weather map is said to be "flat" when this happens. It also is a sure sign that winds are light or nearly calm and vary in direction from place to place and hour to hour. Strong winds, as expected, invariably occur with a strong gradient. As an example, the isobars about a hurricane are extremely close together. More about isobars, gradients, and weather will follow in the discussions of winds, air masses, fronts, and highs and lows. For now it is very important to be aware of the significance of pressure, for together with temperature it is one of the basics for understanding and anticipating weather and weather changes.

Winds

Winds blow because of the differences in atmospheric pressure that exist from place to place; yet air does not blow directly from regions of high pressure to those of low pressure. Rather the air tends to blow parallel to the isobars, so that there is higher pressure to your right when you have your back to the wind in the Northern Hemisphere. (In the Southern Hemisphere, low pressure would be to your right.)

The reason air blows parallel to the isobars rather than directly from high to low pressure is remarkably logical when you take into account the rotation of the earth. Taking it step by step: Air begins by moving away from a high-pressure region directly toward low pressure in much the same way that air from a bursting balloon (high pressure) rushes out to the neighboring air (low pressure). Along its path something changes—not in the air but in the underlying surface. The earth has moved because of its rotation in space. The apparent force exerted by the earth's rotation is called the *Coriolis effect*. The earth rotating beneath the moving air makes the wind appear to curve clockwise until it blows along the isobars. The curvature stops at this point because further turning would require the wind to flow from low to high pressure—like a river trying to flow uphill—which it can't do without the help of another, external force.

This concept of wind flow is very real and is particularly true of air motions high enough above the earth's surface to be free from the effects of ground friction. This type of flow is called *geostrophic.* If the isobars are strongly curved rather than straight, then an added acceleration is present owing to centrifugal force. The wind still parallels the isobars in much the same way, but is instead called a *gradient wind.*

The air close to the ground is slowed down by friction, and the balance between the pressure and Coriolis forces is changed. Because of friction, the air close to the ground both slows down and changes direction by angling toward the region of lower pressure. The rougher the surface, the greater the deflection. Between the ground surface and the "friction-free" atmosphere at the level of the geostrophic wind (up to 3,000 feet), the wind angle may change by 30° over land and 15° over the oceans. At the same time, wind speed is increasing with altitude; the change is most rapid near the ground and tapers off appreciably at 50 to 100 feet.

Another complication observed with surface winds is the way they change between night and day (even if the pressure gradient doesn't). Beginning shortly after sunset and continuing until after sunrise, the wind speed normally drops off dramatically. It's not at all unusual for the nighttime air to be calm. But a look at clouds several thousand feet above reveals that the winds aloft continue to blow.

With the setting sun, the lower layers of the atmosphere cool off more quickly than those aloft. If the surface layers cool sufficiently, temperature will actually increase with altitude. A temperature increase with height, called an *inversion,* signifies that the air is in a very "stable" condition (because the lower, cooler air is much denser or heavier than the air above it). As a consequence the lower air layers become "uncoupled" from the upper layers, and the momentum contained in the wind flow aloft ceases to be transferred to the surface. Without this source of energy

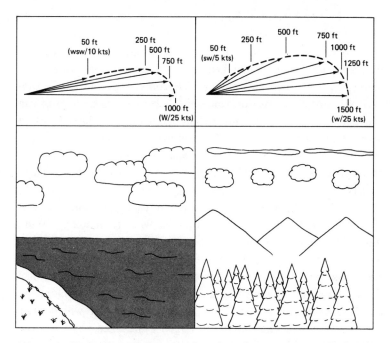

Schematic illustration of the way the wind changes speed and direction with altitude over a large, smooth lake and a rough forest. *The arrows depict the air motion at different heights.*

the winds drop off or even become calm. Actually the air is rarely totally calm because local differences in air pressure or density become important when the large-scale pressure-driven winds die out. These local effects can result in drainage flows where ground elevations vary, or city winds that blow because of the temperature difference between downtown and the outlying suburban and rural areas. But more about these in Part Two.

The opposite situation begins after sunrise when the surface and the overlying air again become warm: Atmospheric mixing and turbulence begin as the lowest layers become warmer than those above. An "unstable" condition results and the wind speed picks up as renewed coupling occurs with the air and winds aloft.

Unsteady, gusty winds with large, rapid changes in speed and direction can result if the heating of the surface is either very intense or sporadic. The military has found that the black, rubberized tarps they use for makeshift helicopter landing pads can produce heating that is intense enough to create dust devils in tropical climates. The sporadic heating of the ground by puffy cumulus clouds that intermittently shade the sun is another cause of varying winds (although the winds are not of the intense, gusty variety).

Having investigated a few facts about winds and their causes, we now need to concern ourselves with wind terminology. Wind direction is given in a way that is contrary to most other directions we are used to dealing with: It is the direction *from* which the air blows. An easterly wind comes out of the east, blows toward the west, and is designated by a compass reading of 090°. Direction can also be expressed by points of the compass, of which there are sixteen. North winds are 000° or 360°, while an east-by-southeast wind is 112.5°. There are a number of ways to report wind speed, ranging from furlongs per fortnight to feet per second, but three are most common: miles per hour, knots (equal to one nautical mile per hour), and meters per second (used mainly in scientific and technical applications, and abroad). To convert from one wind speed system, use the conversions given below.

WIND SPEED CONVERSIONS			
	mps	mph	kt
1 meter per second (mps)	1.00	2.24	1.96
1 mile per hour (mph)	0.45	1.00	0.88
1 knot (kt)	0.51	1.14	1.00

Winds can be measured with three different instruments: A *wind vane* is used to measure direction, an *anemometer* registers speed, and a *propeller vane* measures both speed and direction. Some com-

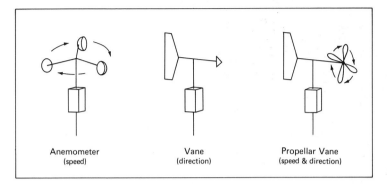

Three basic wind instruments and what they measure

mon wind instruments are pictured above; these are generally used to make routine wind observations at a fixed location: a weather station, on a golf course, or at a vacation home. They are not easily portable. Other, more simple types are designed to be portable and can be a valuable addition to the gear you take with you in the out-of-doors. Appendix D gives some ideas for such a portable weather station. But even if you do not have an anemometer with you, it is still possible to estimate the speed of the wind by the way that leaves blow and trees sway, or the height and character of waves at sea (see Table 4, pages 18 and 19).

Humidity

Have you ever wondered why people sometimes say, "It's too cold to snow"? Even though the saying isn't a particularly accurate aid to weather forecasting, it is based on sound scientific principles. Put another way, the saying is wrong for the right reasons, for it indicates that cold air does not contain very much water vapor. And even if it all turned to snow, it wouldn't amount to very much anyway.

TABLE 4. CHARACTERISTICS AND EFFECTS OF VARYING WIND SPEEDS ON LAND AND AT SEA

Wind Speed (mph)	Wind Force (Beaufort Scale)	Wind Description	Wind Effects and Characteristics	
			On Land	At Sea
<1	0	calm	Smoke rises vertically	Water surface like a mirror
1–3	1	light	Wind direction can be ascertained only by smoke drift	Ripples on the surface, but no foam
4–7	2	light breeze	Wind felt on face; leaves rustle	Small wavelets form; crests look glassy and do not break
8–12	3	gentle (breeze)	Leaves and small twigs in constant motion; wind extends small, light flags	Large wavelets form; crests begin to break—scattered whitecaps
13–18	4	moderate (breeze)	Raises dust and loose paper; small branches move	Small waves become longer; whitecaps more widespread
19–24	5	fresh (breeze)	Small, leafy trees begin to sway; crested wavelets form on inland waters	Moderate waves form, having a pronounced elongated shape; many whitecaps and some spray
25–31	6	strong (breeze)	Large branches sway freely; wires whistle; umbrellas used with difficulty	Large waves begin to form; extensive whitecaps and more spray

Speed	Number	Name	Observed effects on land	Observed effects at sea
32–38	7	high wind	Entire trees in motion; some difficulty walking into wind	Sea "heaps up" and white foam from breaking waves starts to blow in streaks along the direction of the wind
39–46	8	gale	Twigs break off trees; much difficulty walking	Moderately high and longer waves; foam is blown in well-defined streaks
47–54	9	strong gale	Slight structural damage as shingles may be blown loose	High waves; dense streaks of foam blown with the wind; spray affects visibility; sea begins to "roll"
55–63	10	whole gale	Trees uprooted; considerable structural damage	Very high waves; sea surface appears whitish; visibility is impaired; rolling of the sea becomes heavy and shocklike
64–72	11	storm	Rarely experienced over land; widespread damage	Exceptionally high waves; small and medium-sized ships are lost from view for long periods
>72	12	hurricane	Can inflict severe damage and injury from windblown objects	Air filled with foam and spray; sea completely whitish with driving spray; visibility severely restricted; some structural damage to ships

Let's begin thinking about humidity by listing a few basic facts:

- Water can exist in the air either as a liquid, a solid (such as ice or snow), or a gas (water vapor).
- Water vapor is invisible and odorless.
- Cold air cannot hold as much water vapor as warm air.
- It takes less water vapor in cold air to produce snow than is required in warm air to produce rain.

Together, these four basic facts tell us a lot about humidity, the water vapor content of the air. Humidity is an important weather element for at least three reasons: (1) Knowing the humidity of the air and its temperature gives us a good handle on the weather changes that must occur to produce precipitation; (2) humidity is a controlling factor of the comfort we feel in the outdoors; and (3) humidity is a major determinant of visibility—our ability to clearly see the environment around us. We'll get to each of these effects shortly, but first let's touch on two ways of expressing humidity: *relative humidity* and *specific humidity.*

The relative humidity is the amount of water vapor actually present in the air and is expressed as a percentage of the maximum amount of water vapor that the air could hold at that temperature. Since warm air can hold more moisture than cold, air with a relative humidity of, say, 60 percent will contain more water vapor at 80°F (26.7°C) than will 60 percent humid air at 45°F (7.2°C). Stated differently, this temperature dependence is the reason why air with a 30 percent relative humidity at 87°F (30.6°C) would have 100 percent relative humidity at 50°F (10°C). When air has 100 percent relative humidity, it is said to be *saturated;* either adding more water vapor to the air or dropping its temperature will cause the water vapor to condense and form water droplets, ice pellets, or snowflakes. If saturation occurs in air that is in contact with the ground, then dew or frost will form: dew if the temperature of the ground is above freezing (32°F or 0°C), and frost if the ground temperature is below freezing. In the example above, 50°F (10°C) is referred to as the *dew point*—the temperature at which saturation occurred. Had saturation taken place at a temperature below freezing, that temperature would instead be called the *frost point.*

Sometimes humidity is expressed not as the relative amount held by air of a particular temperature, but instead as the absolute amount of moisture present. One way to state humidity in this way is to refer to the specific humidity: the ratio of the weight of the water vapor contained in a sample of air to the total weight (that is, both the moist and dry parts) of the air. A slightly different ratio sometimes used is the *mixing ratio:* the ratio of the weight of the water vapor to the weight of only the dry air. Since the water vapor is only a small fraction of the total weight of the air, specific humidity and mixing ratio are virtually interchangeable for most purposes. At 95°F (35°C) saturated air consists of only about 5.5 percent water vapor (its specific humidity), while its mixing ratio is 5.8 percent.

The simplest way to measure humidity is with a *psychrometer,* a simple device consisting of two thermometers. The two are identical except that one (called a wet-bulb thermometer) has a piece of wet muslin wrapped around its bulb. The muslin is to be kept wet with distilled water, although soft, fresh water will work equally well. Usually the psychrometer is mounted on a sling so it can be twirled rapidly. When this is done, some of the water from the wet bulb evaporates, cooling the bulb in the process in much the same way that our bodies are cooled after stepping out of a shower or when perspiring. *Evaporation* is the process of changing liquid water to water vapor, and it requires an input of heat (whereas *condensation* gives off heat to the environment as the vapor changes to water or ice). The heat for evaporation is supplied to the wet wick by the bulb and to the air in contact with the bulb. The result is that the evaporating water cools the wick and the temperature of the wet-bulb thermometer drops. The difference in temperature between the dry-bulb and the cooled, wet-bulb thermometer is called the "wet-bulb depression" and is a direct indicator of how humid the air is. Saturated air will have a zero wet-bulb depression, while the very dry air in the southwest United States will commonly have depression of 20° to 30°F (11° to 17°C). Once you know the air temperature and the wet-bulb depression, you can use simple tables to determine the relative humidity, dew point, or specific humidity. An example of these is found in Table 5, pages 24 and 25.

Thinking ahead to some specifics that are discussed later in Part Two, we can begin to perceive the importance of knowing what humidity conditions are and how they are apt to change. We can use our knowledge of humidity to judge the chances for fog, dew, or even precipitation. We can also estimate how cold the ground and air are likely to get at night by knowing daytime temperatures and humidities (and carefully observing what clouds are present).

While humidity is a key determinant of temperature and precipitation, it is equally important by itself for the way it directly affects our comfort. I recall that the $-10°F$ $(-23.3°C)$ temperatures I routinely experienced during the dry winter days in Wisconsin were far more bearable out of doors than were the 5° and 10°F days of my Long Island youth. But indoors, things were quite different. Warming Wisconsin's cold, dry air up to 70°F produced the discomforts of dry noses and throats and fingertip spark generators that rivaled Ben Franklin's. True, eastern winters have these effects as well, but not to the extent of the less humid mid- and far-western states.

As mentioned earlier, humidity controls the amount and rate of evaporation and with it the cooling we feel in summer. When the air is dry, perspiration evaporates rapidly, keeping our bodies cool and comfortable. But as the relative humidity increases, the rate of evaporative cooling drops off markedly as the ability of the air to take on more water vapor is diminished. Consequently, 90°F temperatures with 60 percent relative humidity are infinitely more uncomfortable than the 100°F readings common to the southwestern deserts with their 30 percent humidities.

Precipitation

Precipitation includes all forms of water—liquid and solid—that falls to the ground. It is formed when water vapor cools to the dew point or, if the temperature is below freezing, the frost point. Some forms of precipitation we all recognize are drizzle, rain, snow, hail, and sleet.

Drizzle is a fine mist of small liquid droplets that are less than .02 inch in diameter. They fall from stratus clouds. Precipitation from drizzle does not collect on the ground at a rate of more than .04 inch per hour.

Rain is drizzle's big brother and includes all liquid "precip" greater than .02 inch in diameter. On the other end of the scale, raindrops are usually not larger than .2 inch in diameter because they tend to break up into smaller drops. Rain is *light* if it falls at a rate of less than .10 inch per hour, *moderate* if it falls at a rate of up to .30 inch per hour, and *heavy* if it falls at more than .30 inch per hour. The world's heaviest rainfall was 12 inches in forty-two minutes at Holt, Missouri, on June 22, 1947.

Snow consists of clear or cloudy ice particles in the form of individual *crystals,* small *pellets,* or *flakes.* Flakes are caused by the gathering together of crystals; they are the most common form of snow and usually range in size from .1 inch in cold weather up to an inch or so near freezing. Snowflakes may cluster until they reach 3 or 4 inches in diameter. In very still air, flakes up to 10 inches in diameter have been reported. Snowfall is reported either as the actual depth of the snow or as the liquid water equivalent; a rough average is that the depth is ten times the liquid equivalent. The greatest snowfall accumulation from a single storm was 189 inches at Mt. Shasta, California, over a period of seven days in February 1959.

Hail is a lump of ice that is roundish or irregular. Hailstones are usually about .4 inch in diameter, but in the Midwest they can reach 4 inches. The largest hailstone recorded was the size

TABLE 5. RELATIVE HUMIDITY (%) AND DEW-POINT TEMPERATURE

Air Temp. (°F)	Difference Between Air Temperature and Wet-Bulb Temperature (°F)										
	1	2	3	4	5	6	7	8	9	10	11
20	85	70	55	40	26	12					
	16	12	8	2	-7	-21					
25	87	74	62	49	37	25	13	1			
	22	19	15	10	4	-3	-9	-15			
30	89	78	67	56	46	32	26	16			
	27	25	21	18	15	8	1	-7			
35	91	81	72	63	54	45	36	27	19	10	
	33	30	28	25	22	17	13	7	0	-11	
40	92	83	75	68	60	52	45	37	29	22	15
	38	35	33	30	28	25	21	18	14	7	-1
45	93	86	78	71	64	57	51	44	38	31	25
	43	41	38	36	34	31	28	25	22	18	13
50	93	87	80	74	67	61	55	49	43	38	32
	48	46	44	42	40	37	35	32	29	26	22
55	94	88	82	76	70	65	59	54	49	43	38
	53	51	50	48	46	43	41	38	36	33	30
60	94	89	83	78	73	68	63	58	53	48	43
	58	57	55	53	51	49	47	45	43	40	38
65	95	90	85	80	75	70	66	61	56	52	48
	63	62	60	59	57	55	53	51	49	47	45
70	95	90	86	81	77	72	68	64	59	55	51
	69	67	65	64	63	61	59	57	55	53	51
75	96	91	86	82	78	74	70	66	62	58	54
	74	72	71	69	68	66	65	63	61	59	57
80	96	91	87	83	79	75	72	68	64	61	57
	79	77	76	74	73	72	70	68	67	65	64
85	96	92	88	84	81	77	73	70	66	63	59
	84	82	81	80	79	77	76	74	73	71	70
90	96	92	89	85	81	78	74	71	68	65	61
	89	87	86	85	84	82	81	79	78	76	75
95	96	93	89	86	82	79	76	73	69	66	63
	94	93	91	90	89	87	86	85	84	82	81
100	96	93	89	86	83	80	77	73	70	68	65
	99	98	96	95	94	93	92	90	89	87	86

HOW TO USE THE TABLE

The table can be used to determine either the relative humidity of the air or the temperature at which dew (or frost) will form. First you need to know both the temperature of the air and the wet-bulb temperature (from a psychrometer, or wet-bulb thermometer). Next subtract the wet-bulb temperature from the air temperature to get the wet-bulb depression,

12	13	14	15	16	17	18	19	20	25
7									
4									
8	**12**	**6**							
7	*−3*	*−14*							
7	**21**	**16**	**10**	**5**					
8	*13*	*8*	*0*	*−13*					
3	**28**	**23**	**19**	**11**	**9**	**5**			
7	*24*	*20*	*15*	*9*	*0*	*−12*			
9	**34**	**30**	**26**	**21**	**17**	**13**	**9**	**5**	
5	*32*	*29*	*25*	*21*	*16*	*11*	*3*	*−8*	
4	**39**	**35**	**31**	**27**	**24**	**20**	**16**	**12**	
2	*40*	*37*	*34*	*31*	*27*	*24*	*19*	*14*	
8	**44**	**40**	**36**	**33**	**29**	**25**	**22**	**19**	**3**
9	*47*	*44*	*42*	*39*	*36*	*33*	*30*	*26*	*−11*
1	**47**	**44**	**40**	**37**	**34**	**30**	**27**	**24**	**9**
5	*53*	*51*	*49*	*47*	*45*	*42*	*39*	*36*	*15*
4	**50**	**47**	**44**	**41**	**38**	**35**	**32**	**29**	**15**
2	*60*	*58*	*56*	*54*	*52*	*50*	*47*	*44*	*28*
7	**53**	**50**	**47**	**44**	**41**	**38**	**36**	**33**	**20**
8	*66*	*64*	*63*	*61*	*59*	*57*	*55*	*52*	*39*
8	**55**	**52**	**49**	**47**	**44**	**41**	**39**	**36**	**24**
3	*72*	*70*	*69*	*67*	*65*	*63*	*61*	*59*	*48*
1	**58**	**55**	**52**	**50**	**47**	**44**	**42**	**39**	**28**
9	*78*	*76*	*75*	*73*	*72*	*70*	*68*	*66*	*56*
2	**59**	**56**	**54**	**51**	**49**	**46**	**44**	**41**	**30**
5	*84*	*82*	*81*	*79*	*78*	*76*	*74*	*72*	*63*

and locate the vertical column that has this value. Then locate the row with the correct air-temperature value. Where the row and column intersect you will find two numbers: the upper value (in boldface) is the relative humidity; the lower value is the dew point (in italics). Example: If the air temperature is 70°F and the wet-bulb temperature is 63°F, then the relative humidity is 68% and the dew-point temperature is 59°F.

of a grapefruit and weighed 1.5 pounds; it fell on July 6, 1928, in Potter, Nebraska. Hail forms in cumulonimbus (storm) clouds where layers of water freeze on the surface of the hailstone during repeated ascents and descents in the updrafts and downdrafts within the cloud. The hailstone finally falls out of the cloud when its weight is too great to be supported by the updrafts.

Sleet or ice pellets are small transparent or translucent particles of ice that form when liquid droplets freeze as they pass through a layer of cold air.

Glaze is not a true form of precipitation but is a consequence of cold rain landing on a subfreezing ground surface. The result is that the water freezes on impact, coating everything it touches with an ever-thickening layer of ice.

Visibility

Unlike temperature, pressure, or humidity, visibility is not a true element of the air itself. Rather it depends on two major factors—one atmospheric, the other human. The nature or composition of the air is one determinant of visibility and the ways it affects the propagation of light. The way in which we humans perceive those light waves reaching our eyes is the second.

The lack of adequate visibility can put us in an unsafe situation while boating, flying, skiing, or hiking; poor visibility can indicate levels of man-made pollution that are dangerous to our health; and we can become depressed and irritable when reduced visibilities persist for extended periods. When the visibility is excellent, we are able to enjoy to unparalleled intensity the many beautiful vistas and panoramas nature provides. But how many visitors to Los Angeles are unaware that the city is encircled on three sides by beautiful mountains? Without visibilities of at least 50 miles, the Grand Canyon would not be worth viewing. Imagine the disappointment when the Rockies cannot be seen from Denver, or when they take on the ugly hue of the "brown cloud" that is Denver's

WEATHER EXTREMES IN NORTH AMERICA

1 U.S. lowest avg. annual temp.: **10°F**
   U.S. coolest summer avg.: temp.: **37°F**

2 U.S. lowest temp.: **-76°F**

3 Alaska's greatest avg. annual precip.: **220 in.**

4 Conterminous U.S. greatest avg. annual precip.: **144 in.**

5 U.S. foggiest place: **2552 hr./yr. avg.**

6 U.S. greatest 24-hr. temp. fall: **100°F**

7 U.S. lowest temp. (excl. Alaska): **-70°F**

8 North America's greatest snowfall in one storm: **189 in.**

9 U.S. greatest depth of snow on ground: **451 in.**

10 Western Hemisphere's highest temp.: **134°F**
   U.S. lowest avg. annual precip.: **1.63 in.**

11 North America's lowest avg. annual precip.: **1.2 in.**

12 U.S. longest dry period: **767 days**

13 North America's greatest 24-hr. snowfall: **76 in.**

14 World's largest officially recorded hailstone: **1½ lb.**

15 U.S. greatest 2-min. temp. rise: **49°F (from -4°F to 45°F)**

16 U.S. highest avg. annual temp.: **77°F**
   U.S. warmest winter avg.: **70°F**

17 World's greatest 42-min. rainfall: **12 in.**

18 World's greatest one-min. rainfall: **1.23 in.**

19 Eastern U.S. foggiest place: **1580 hr./yr. avg.**

20 World's highest surface wind: **231 mph**
   U.S. highest avg. annual wind speed: **35 mph**

21 Greenland's lowest temp.: **-87°F**

22 N. America's lowest temp.: **-81°F**

23 U.S. coldest winter avg. temp.: **-16°F**

answer to Los Angeles (and Houston, Chicago, and the Washington-Boston megalopolis). Yes, visibility is an important aspect of our atmospheric environment and our national heritage that we value beyond conscious awareness. Without good-to-excellent visibilities, most outdoors-oriented people would suffer a major loss.

Up to now, we have spoken of visibility and its impairment as if the terms were precise and their meanings identical to everyone. Obviously, neither is true, not only among laymen, but also among scientists. Visibility implies three concepts: *range, contrast,* and *color.* When visibility is impaired, any one aspect (but usually all three) can be affected. The distance at which we are able to recognize large objects may be reduced, our perception of the sharpness or detail or the features of the object may be impaired, or the color of the object itself may be altered. The ability to detect these changes varies from person to person, as does the impact and significance of such changes.

Visibility impairment results when gases and small particles present in the air absorb and scatter light that is directed to our eyes from some object we are viewing. If the gases and particles absorb the light, the brightness of the object will be reduced or its color altered. The amount of absorption depends on the nature and abundance of the gases and particles that are present in the air and the distance between the viewer and the object. Absorption of light by particles is usually more important than by gases. Particles containing carbon are the most efficient types contributing to the reduction of visibility; these particles usually are emitted by the external combustion of coal and petroleum and from internal-combustion diesel engines. The only significant light-absorbing gas is nitrogen dioxide (NO_2), which has its origins in the exhausts of motor vehicles and the smokestacks of power plants, smelters, and other high-temperature processes using oil or natural gas. NO_2 absorbs primarily blue light and as a consequence will often give the sky a brownish hue.

Whereas light absorption by particles and gases can reduce visibilities perceptibly, scattering effects are usually far more pronounced. To understand how light scattering can impair visibility, we first need to understand how we view an object. All objects that we are able to see direct light rays to our eyes; while some objects emit their own light, most reflect light from some source like the sun. If all of the light from the object were to move in a direct line to us, we would see a perfectly clear and detailed image. But when fine particles or air molecules exist in the air between us and the object, two things happen: (1) Some of the "object light" is deflected out of our field of vision, and (2) some extraneous light (for example, from the sky) is deflected into our sight. What we now see as the object is actually some combination of light from the object and extraneous light. As the number of light scatterers increases, the object appears less clear and detailed.

As with absorption, light can be scattered by gases as well as particles, although small particles are more significant. For example, visibility in the presence of only light-scattering air molecules (that is, air without NO_2 and particles) will be "restricted" to no more than 200 miles at sea level and about 250 miles at an elevation of 10,000 feet. A more important aspect of light scattering by air molecules is its role in creating blue skies and red sunsets. Air molecules scatter the shorter blue wavelengths more than the longer red wavelengths, with the result that the blue portion of the sun's rays are scattered out of the direct path to the sun and into the surrounding sky. Without the blue rays, the sun appears reddish. At sunset, the sun's rays pass through a deeper layer of atmosphere than when the sun is overhead at noon; correspondingly, more scattering occurs at this time and the reddish color is the result.

Particles of many sizes are found in the air. Some are put there naturally, others are man's contribution, while still others are actually formed in the atmosphere as the result of complex chemical reactions between different gases. In turn, these particle-producing gases can have widely different origins, such as the smog layers that often blanket our major cities or the terpenes emitted from pine needles in forests of the Pacific Northwest, the Blue Ridge Mountains, and the Great Smokies. The last two forests may actually owe their names to their terpene production and subsequent visibility effects, although Indian and pioneer campfires may have contributed greater amounts of the light-scattering small particles that gave these areas their bluish haze.

Visibility impairment in the Great Smoky Mountains. *The upper photograph is a clear day with a visual range of 50 miles; the lower photograph is of the same vista but with a much lower visual range (15 miles).* Courtesy of Douglas Latimer, "Relationships Between Air Quality and Human Perception of Scenic Areas," American Petroleum Institute Report 4323, 1980

The particles that have the greatest effect on our ability to see objects clearly over long distances are those that are among the smallest found in the air. If we picture them as spherical, then the diameters of the most effective light-scattering particles (or aerosols, as the smaller particles are often called) are one-millionth of a meter (called a micrometer and abbreviated μm) and smaller. To picture how small these particles are, think that it would take 1,000 of them arranged in a straight line to equal a spread of 1 millimeter (mm), the smallest division found on the common metric ruler. Because they are so small and effective, they do double damage when you consider that on a pound-for-pound basis there are on the order of 15,000 small particles for each large particle.* The situation is further compounded because the smaller and lighter particles float in the air like a gas until they are washed out by rain, snow, or other forms of precipitation. But the larger and heavier particles are unable to remain suspended and will fall to the ground in a matter of hours.

When visibility becomes impaired, it is the result of one or more of the various scattering and absorption processes. In all, there are six categories of visibility impairment that occur commonly. Some are localized and others are widespread, and natural effects are to blame in some cases while the remainder are caused by humans.

1. *Localized, low-humidity smog* is one of the most infamous of the six categories. Los Angeles and Denver have two of the better-known examples. Low visibility conditions are restricted geographically by mountains that inhibit the horizontal movement of the particle-laden air, while an elevated inversion acts as a lid to cap the polluted layer. Visual ranges as low as 1 to 3 miles (2 to 5 km) can result along with discoloration from the presence of NO_2.

---

*The atmosphere contains two "families" of particle sizes. The smaller ones range from .2 to 1 μm, the most common diameter being .3 μm. The larger particles typically range in diameter from 5 to 30 μm.

2. *Widespread haze* can envelop several states at a time (up to 600 to 1,200 miles) during the summer in the midwestern, eastern, and southeastern portions of the country. Humidities are frequently only low to moderate, yet visibilities are reduced to 5 miles (8 km) or less on the average. These episodes are typically caused by a low-wind, high-pressure system that stagnates east of the Mississippi. As a result, pollution sources over many states feed their emissions into the same mass of air for days on end, resulting in a steady production of particles in the atmosphere and a corresponding impairment of visibility.

3. *Smoke plumes* are the consequence of gaseous and particle emissions from single large factories, power plants, smelters, or refineries. Their effects are usually localized, but can be severe if highways, scenic vistas, or population centers are located nearby. The problem can become particularly intense if atmospheric humidities are high, as the moisture emitted in power plant or pulp mill plumes (to name only two) can condense to cause dense fog.

4. *Long-distance transport* of smoke or gaseous plumes can create noticeable reductions in visibility, particularly in the pristine portions of the Southwest. Perhaps most important is the aesthetic impact on the majestic vistas that are an integral part of that region's identity and character.

5. *Windblown dust* is a low-visibility category predominantly of natural origin. Although sand and dust are some of the larger particles found in the air, the large amounts picked up by the wind can lead to visibilities less than one-half mile in dust storms.

6. *Fog and high-humidity haze* comprise one of the poorest visibility categories. Fog may be naturally produced or may be triggered by water vapor emissions from industrial sources. Visibilities less than 100 meters are not uncommon. When relative humidities exceed 95 percent, very small particles will grow in size as they take on water from the surrounding air, and high-humidity haze can form. The effect is similar to a fog and is often reported as such.

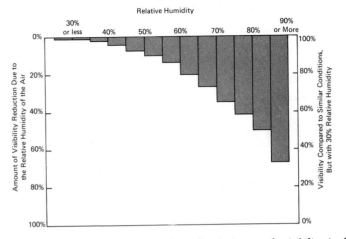

Effect of humidity on visibility. *As humidity is increased, visibility is decreased. The illustration assumes a "reference" visibility exists when the air has 30% relative humidity. As the humidity increases, the ratio of the actual visibility to the reference decreases as shown. For example, if 30% humid air had a 50-mile visibility, then the visibility would decrease to one-half or 25 miles if the humidity increased to the 80–85% range.*

In our discussion of visibility and its impairment, effects due to humidity have been carefully worded. From your own experiences and recollections, you have undoubtedly observed a strong correlation between increasing humidity and decreasing visibility. This relationship is indeed substantiated by scientific observations in spite of the fact that water vapor itself neither scatters light rays nor absorbs them. The visibility effects attributable to humidity are due to the way in which certain types of particles "grow" in size in humid air, much the same way that a sponge expands as it absorbs water. These particles are called *hygroscopic* and may be of either natural or man-made origin. Sea salt is one type of hygroscopic particle, and you can easily witness its affinity for water vapor by noting how visibility at the shore quickly responds to changes in the relative humidity. When the midday humidity is sufficiently high (above 70 percent), you will often be able to notice the decrease in visibility in the late afternoon as the air temperature drops, the relative humidity increases, and these particles grow and increase their light scattering.

Chapter Two
Workings of the Atmospheric Weather Machine

Understanding and predicting the weather—particularly conditions arising from local or so-called micrometeorological effects—has many similarities to the way in which a mechanic goes about diagnosing and repairing your car's engine. If all you know is that depressing the gas pedal makes the iron machine go faster, then you are apt to have a difficult time locating and repairing a defective solenoid. To get a handle on your vehicle's problem, you need to have some understanding of its electrical, fuel, lubricating, and coolant systems. Without this knowledge you may occasionally stumble on the source of the problem, although the odds will not be in your favor. Like the engine in your car, the atmosphere is also a heat engine or weather machine. It has its fuel sources and its coolant and lubricating systems, but they are less easily recognized and understood. To make more accurate forecasts, particularly of local weather conditions, and to understand how weather is created in the first place demands an understanding of the sun, the earth, and the ground and oceans, and the way in which they interact.

The Sun—Earth's Ultimate Energy Source

The source of all energy contained in our atmosphere and on our planet is the sun. Whether this energy is in the form of electricity, wind motion, heat, or thermal radiation, its origins can be traced to the sun. Separated from earth by an average distance of 93 million miles (150 million km), the sun generates about one million billion times more energy than the capacity of all electric-generating power plants in the United States. The sun consists mainly of hydrogen (about 75 percent by mass) and helium (about 25 percent), and measures about 865,000 miles in diameter—about 109 times the earth's diameter. The sun is a large thermonuclear furnace that changes hydrogen into helium, giving off heat, light, and other energy forms in the process.

The sun is responsible not only for the energy the earth receives during the daylight hours, but also for almost all other energy forms we use, except nuclear power. Hydroelectric energy has its origins in the storms that deposit rain and snow at the higher elevations; the energy that initially evaporated the moisture from lakes, rivers, and oceans was provided by the sun. Wind power is another clear-cut example of solar energy that is captured in a different form. But fossil fuels as well had their energy supplied by the sun. Plants convert the sun's energy in photosynthesis to grow; over millions of years, some of these plants have been transformed into the fossil fuels that we burn today. Other conversions haven't taken as long, like the wood we use as our nearly exclusive energy source in the outdoors and as an ever-popular indoors energy source as well.

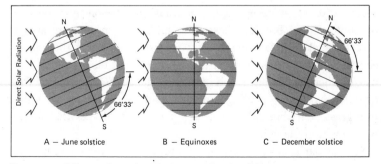

Exposure of the earth to the midday sun for the different seasons

Energy travels from sun to earth in two forms: *electromagnetic radiation* and *corpuscular radiation.* Corpuscular radiation consists of alpha rays (a stream of positively charged particles), beta and cathode rays (high-energy electrons), and neutrons and protons. Corpuscular radiation has no significant effect on weather processes, although it affects us in two other ways: It causes radio interference, and it produces the aurora borealis (Northern Lights) and aurora australis (Southern Lights). These are beautiful, natural light shows that result when protons and neutrons emitted from the sun interact with oxygen and nitrogen in the upper reaches of the earth's atmosphere (at a height of around 40 miles, or 70 km, in a region called the thermosphere). The corpuscular radiation transfers its energy to oxygen atoms and molecules and to nitrogen molecules, placing them in a highly energized state. These atoms and molecules then release this excess energy in the form of visible light as they return to their normal energy level. Corpuscular radiation is attracted to the earth's magnetic poles, and so the Northern Lights are usually seen only in Canada, Alaska, and the northern states. The best months for seeing the aurora are March–April and September–October. Auroral displays are particularly frequent during years of maximum sunspot frequency, every eleven years (1980, 1991, and so forth) when they can also be seen at more southerly latitudes in the Northern Hemisphere.

The aurora are truly one of nature's magnificent productions. Most of us have never seen this transient phenomenon outside of photographs. But for those who have, the spectacular arrays of color and form are not readily forgotten. Green is their most common color, although reds, pinkish hues, and blues are not unusual. Their appearance and position change perhaps more frequently than their shading. Some extend for hundreds of miles in a great arc, while others resemble curtains or drapes that dance like chimes in the wind.

Spectacular as may be the visual effects of corpuscular radiation, so-called electromagnetic radiation accounts for 99.9999 percent

TABLE 6. THE ELECTROMAGNETIC SPECTRUM	
Types of Electromagnetic Radiation	Wavelengths
Long radio waves	Longer than 2,000 meters[1]
Standard AM radio broadcasts	100 to 2,000 meters
Short radio waves	20 centimeters to 100 meters[2]
Microwaves	.1 millimeter to 20 centimeters[3]
Infrared radiation	7 ten-millionths meter to .2 millimeter[4]
Visible light	4 to 7 ten-millionths meter
Ultraviolet radiation	1 hundred-millionth to 4 ten-millionths meter
X rays	1 ten-billionth meter to 2 hundred-millionth meter[5]
Gamma rays	Shorter than 1 ten-billionth meter

[1] 1 meter = 39.372 inches
[2] 1 meter = 100 centimeters; 1 centimeter = .3937 inch
[3] 1 centimeter = 10 millimeters
[4] 1 millionth of 1 meter is called a micrometer (μm)
[5] 1 ten-billionth of 1 meter is called an Ångstrom unit (Å).

of the sun's energy that reaches the earth. Electromagnetic radiation includes a broad spectrum of familiar energy forms: radio waves, microwaves, infrared radiation, visible light, ultraviolet radiation, and X rays and gamma rays. Electromagnetic radiation can be a difficult concept to comprehend. While all forms have different energy levels and varied effects, they share certain common features. They have the properties both of waves and particles. These extremely small energy particles are called photons and have a finite mass and an energy level that are unique to the individual form of electromagnetic radiation. Each photon has both an electric and a magnetic field associated with it. These fields, in turn, also have the characteristic of waves. As a comparison, let's consider for the moment sound waves—which are *not* a form of electromagnetic radiation but which do contain and transmit energy. When you feel the vibrations of a speaker playing the music of a bass drum, you know that its frequency is low (conversely, a long wavelength) compared with the frequency of a piccolo (which has shorter wavelengths). In an analogous way, the various forms of electromagnetic radiation have different wavelengths, frequencies, and amplitudes (or energy levels).

The two forms of electromagnetic radiation that play key roles in understanding and making weather are *visible radiation* (light) and *infrared radiation*. All forms of light have higher frequencies (shorter wavelengths) than do all segments of the infrared portion of the electromagnetic spectrum. Actually, these two energy forms are neighbors on the electromagnetic spectrum: The low-frequency end of the visible portion is seen by the eye as red light, which gradually becomes invisible as the frequency on the spectrum decreases to the point where it blends into the infrared portion of the spectrum.

The majority of the electromagnetic radiation received by earth from the sun is in the visible portion of the spectrum. It is no accident that the human eye is most sensitive to this form of electromagnetic radiation, having its peak sensitivity at the green wavelengths. The type of radiation given off by an object (not to be confused with energy that is reflected by an object) is controlled by its temperature. The exterior of the sun has a surface temperature of about 6,000°K,* while the earth has an average surface temperature around 284°K (51°F) when all water, land, vegetation, and ice surfaces are averaged. At this temperature, the earth radiates electromagnetic radiation in the form of infrared radiation (also called longwave radiation, as distinguished from the shortwave radiation of the sun). The atmosphere also radiates infrared energy consistent with its global-average temperature of 250°K.

---

*Absolute temperatures (°K) have their zero point where molecules cease to move. A change in temperature of 1°K equals a 1°C change. In the absolute system, water freezes at 273°K and boils at 373°K.

Auroral displays
at Fairbanks,
Alaska. *The
three forms
shown are
multiple bands*
(top), *corona*
(center), *and
rayed arc*
(bottom).
Courtesy of
Jack Finch

Energy Processes on Earth

There are six ways or processes in which the energy from the sun is used. One of these, infrared radiation, has already been discussed. The remaining five are (1) *heat conduction in the ground,* (2) *heat conduction in the air,* (3) *evaporation,* (4) *condensation,* and (5) *photosynthesis.* It is important to be aware of each because of the effects they can have on the local weather of individual locations and the weather patterns they can create or change over large regions. On the very local scale it is the distribution of the available solar energy that makes us comfortable at certain times or places and uncomfortable at others. If a large percentage of the sun's energy is captured and retained by our local environment, then there may be sufficient heat in the air and ground to provide us with comfort. Or there may be too much, in which case we need to avoid or reject some to avoid heat prostration or just plain discomfort. Obviously, insufficient heat can work in the opposite direction to make us uncomfortable in other ways.

Heat conduction into or out of the ground is an important energy process. The heat contained in the earth's molten core is isolated from the surface for all practical purposes. In a few cases, to be sure, this heat does make its way upward to warm the ground and the overlying layers of air. Geysers and volcanoes are two geographical features that transfer this energy to the surface, but while this heat source is important in geothermally active areas, its overall significance is negligible.

The heat that is contained in the upper layers of the ground was originally provided by the sun. The uppermost layer of the ground exposed to the sun absorbs a fraction of the sun's energy impinging on it. Dark, rough surfaces absorb more of this available solar energy than do smooth, light-colored surfaces. The fraction of solar energy not absorbed at the surface is reflected skyward and is called the *albedo.* The amount actually retained serves to warm the ground (or water) or evaporate water (we'll discuss the evaporation process shortly). Very dark, dense surfaces absorb the sun's energy in a very shallow layer that is less than a millimeter thick. Water and ice are translucent to the sun's light, and the energy is absorbed over a correspondingly thicker layer. In clear water, about 75 percent of the energy contained in the blue and green wavelengths penetrates to a depth of about 33 feet (10 m), while 6 percent is still present at 330 feet (100 m). The longer red wavelengths do not penetrate as far: Only 36 percent penetrates to a depth of 3 feet (1 m), while only 2.6 percent reaches 33 feet (10 m). The rapid absorption of the orange and red light in clear water is responsible for the blue-green coloring of most objects at the bottom. Two changes occur when the water is murky: The depth to which light can penetrate is reduced, and the maximum penetration shifts to the orange and red wavelengths. As a consequence, those objects that can be seen are more apt to take on yellow, orange, or red hues. But relative to clear water, the penetration depth for all wavelengths is markedly reduced; it is just that the reduction is most pronounced for the violets, blues, and greens, and helps to explain the coloration of dirty water.

Light has difficulty in penetrating both snow and ice, compared with lake water. Three-fourths of the incoming light is absorbed in less than 1 foot (25 cm) of glacial ice, and 75 percent of the sun's energy is absorbed in a few inches (10 cm) in snow.

Heat conduction into or out of the underlying surface depends on three factors: (1) the depth at which the heat is stored, (2) the temperature difference between the heat source and the adjacent depths, and (3) the ability of the soil or water to conduct the heat. For most soils, the sun's energy is absorbed at the surface. The temperature reached at the surface depends on the composition of the soil, in addition to the influence exerted by the other energy processes. The conduction of heat under the ground surface depends on the type of soil, rocks, and minerals as well as the moisture present in the ground. Clay is a good conductor of heat, water is five times poorer, while humus conducts heat ten times more poorly and air about a hundred times. Said another way, soil that has been aerated or that contains a large fraction of organic matter is a good insulator, and soil that is wet or has a high clay content is a good heat conductor. A humus mulch exposed to the sun will heat rapidly during the daytime, although the high air and organic content will confine the heat to a shallow layer at the surface. The insulating properties of the mulch will discourage conduction of the heat to the underlying soil layers; this will be partly offset by the large temperature difference between soil and mulch surface. At night the mulch surface will cool as it gives off infrared radiation (it is doing this in the daytime as well, but the infrared cooling is overshadowed by the sun's heating). The underlying soil and plant roots are protected, however, because the poor conductivity of the mulch retards the upward conduction of heat away from the (now) warmer soil.

Heat conduction between the surface and the underlying soil does not occur instantaneously. In the case of the organic mulch, there is a three-hour delay between the maximum daytime surface temperature and the peak temperature at a depth of $1\frac{5}{8}$ inch (4.1 cm), while in ice the same three-hour delay has the heat penetrating three and one-half times more deeply (14.3 cm). The penetration in stirred water is more pronounced where a three-hour lag is found between peak temperatures at the surface and those at 16 feet (5 m). With this rapid transfer of heat, the surface requires a long time to heat, as witnessed by the long lag between sunbathing weather and swimming weather.

As winter changes to spring and then summer, the average daily temperature of the ground surface gradually increases. This is a reflection of the annual march of the seasons. Just as the ground responds to the daytime-nighttime heating and cooling cycle, it also responds to the annual cycle. When the ground surface reaches its peak daily-average temperature for the year, there is a two-month wait until the 3-foot (1-m) depth of an organic soil reaches its daily-average peak temperature. Clay and ice respond similarly; a two-month lag in peak daily-average temperatures is found at the 11-foot (3.6-m) depth. This again indicates the wide differences in the heat conduction characteristics of various commonly found surfaces.

Very little of the sun's energy heats the atmosphere directly. The amount is so small that it can be ignored for almost all practical considerations. One extremely important process by which the air is warmed (or cooled) is by conduction. In some ways heat conduction in the air is very similar to conduction in soil. In other ways it is quite different. Heat is conducted from warm air to cooler air; the amount of heat transferred depends on the difference in temperature between two locations. A large temperature difference indicates a *potential* for significant heat transfer from the high-temperature region to the cooler region. No temperature difference, no heat conduction. The conductivity of the air, unlike that of soil, does not depend on its composition. Its water vapor, carbon dioxide, or ozone concentrations have no practical effect on heat conduction. But the conductivity of the air, like that of water, is influenced by its movement, its turbulent motions. Calm air is a poor conductor of heat, while the air on hot and unstable summer days is an excellent conductor.

Heat conduction in the air is usually thought of in two different ways. One process, *advection,* transports heat horizontally from warmer regions to colder ones. A vivid example of advection is the heat transported northeastward in summer from the Gulf of Mexico into the middle eastern and northeastern states. There can be no question that large quantities of heat are being transported through the air. This kind of large-scale transport is associated with large expanses of air that are called air masses (discussed separately in Chapter 4). Air transported from one region to another in this way usually completely overwhelms local weather factors, and even when it does not do so totally, it is sure to modify local conditions significantly.

The second process, the conduction or transport of heat between overlying air layers or between air and ground, is one of the most important phenomena controlling local weather conditions. As the sun heats the ground surface, its temperature rises until it may become warmer than the overlying air. When this happens, the heat from the ground will be conducted into and through the air. As the wind blows harder, the air is mixed more rapidly and the transfer of heat from ground to air speeds up. As heat leaves the ground, the temperature of the ground will increase less rapidly or the ground may cool off even though the sun continues to shine. This is what happens after midafternoon on clear days when the ground temperature begins to fall although the air temperature may continue to rise.

Wind motion is an effective atmospheric mixer, enhancing the transfer of heat. Without air movement (and the turbulence it creates) heat transfer by conduction is virtually cut off—except for an inconsequential amount due to the movement of the gas molecules that comprise the atmosphere. But on clear, sunny days, effective heat conduction is possible even though the horizontal winds may be light or calm. This is especially possible over dark, dry, sandy areas that absorb much of the sun's energy. When the ground surface and the lowest layers of air in contact with it (particularly the lowest few feet or tens of feet in exceptional cases) heat up rapidly, the density of the air in contact with the ground decreases sharply in comparison with the now-cooler and heavier overlying air. The effect is very similar to that of a helium-filled balloon or warm smoke from a campfire. The warm bottom air rises until it encounters air at some height with density equal to its own. While it is rising, it exchanges heat with the surrounding cooler air and so it actually becomes denser along its upward path. It is possible to calculate the temperature difference that must be attained for this vertical conduction—or *free convection,* as it is called—to take place of its own impetus. If the temperature of dry air decreases at a rate greater than about 1°F for every 50 feet of altitude (3.4°C per 100 m), then free convection should theoretically begin. Other factors at work in the atmosphere actually inhibit free convection until greater temperature gradients are attained. However, the temperature variation near the surface can be so large on most sunny summer days that limited free convection is a common phenomenon. You can observe its presence by viewing the horizon with your head within a foot or two of the ground; the dancing of distant objects results from the density variations in the air as it alternately bubbles away from the surface.

Heat conduction in the atmosphere does not always extract heat from the ground or transfer it from lower to higher altitudes. In a similar way heat is regularly transferred downward as well, often being a source of warmth for the ground. This is the situation that routinely prevails during clear nights; the ground surface cools off rapidly by emitting infrared radiation and establishing a temperature inversion in the process. The rate of cooling of the ground is decreased and can ultimately be stopped in time as the heat from the overlying, warm air is conducted downward. The rate of heat transfer unfortunately is usually slowed because of the light winds that are typical of nighttime.

More dramatic examples of heat conduction toward the ground are often associated with advective conditions. Spring thaws are often hurried along as warm air is transported northward. Extremely strong inversions result as the warm air blankets the still-frozen ground of the upper Midwest, conducting its thermal energy downward and often melting the snow pack faster than the natural stream-river drainage system can handle.

Evaporation and condensation can dominate the other energy-transfer processes at certain times; at other times their influence is negligible. Evaporation occurs when liquid water, snow, or ice uses energy to change to gaseous water vapor; the energy for evaporation is usually supplied directly by the sun or indirectly by heat conduction from either the air or the ground. Because energy is required to evaporate water, the source of the energy consequently gives up some of its energy or heat and cools in the process. Perhaps the most common example of this effect is the cooling sensation you feel after stepping out of the shower. The water on your skin takes heat from the surrounding air and from your body, lowering your skin temperature while it evaporates. For each gram of water that evaporates, an input of nearly 600 calories is required. Historically, the continental United States receives a daily average of 500 to 750 calories of energy from the sun over each square centimeter of ground surface during a typical July day. If all of this available energy were used only for evaporating water, we could quickly estimate how large a heat sink evaporation can be. Since a cubic centimeter of pure water weighs 1 gram, a typical July day's solar energy input could evaporate a layer of water only about 1 cm ($\frac{3}{8}$ inch) deep—and then only if none of the energy was used to heat the air or ground. Since snow on the average is one-tenth as dense as liquid water, the depth of snow evaporated might be about 10 cm, except that the typical January day receives only 100 to 200 calories per square centimeter in the snowy regions. However, mountain slopes that face the sun will get appreciably more of the sun's energy, and those that face northward will get much less. One further complication with snow and ice: Because they are frozen, they need more energy (about 10 percent more) in order to evaporate.

Condensation is the opposite of evaporation, and since evaporation is an energy drain, it follows that condensation is a source of heat. The concept is not too different from that of a flashlight: The battery retains stored energy until a switch is thrown, at which time the bulb radiates electromagnetic energy (light) and gives off heat by conduction. When a cloud forms in the atmosphere, water or ice droplets are formed when water vapor gives up its stored (or "latent") energy during the condensation process. Why then does a summer shower cool us off on a hot day if the condensation process that formed the droplets actually warmed the surrounding air? The answer is twofold: First, we feel cool when the falling raindrops drag cooler, high-altitude air down with them; and second, the rain that has fallen on hot pavements and warm bodies begins to evaporate, cooling them off in the process.

Another energy-transfer process is photosynthesis. This is the means by which light is absorbed by chlorophyll in the leaves of plants. The absorbed energy causes carbon dioxide to react with hydrogen atoms in the water contained in the plant to produce sugar, giving off oxygen to the air as well. The total amount of energy used in photosynthesis is not particularly large (about 5 percent of the available solar energy), although it is an obviously vital usage. It is in essence this energy that feeds our own bodies by providing plant food for direct consumption and fodder for livestock. Photosynthetic energy is also used later (sometimes much later) by man in the combustion of wood and fossil fuels and the production of biofuels.

Other energy processes can be important at particular times and locations, although they may be nearly immeasurable when averaged on a worldwide basis. Melting snow and ice are one common example. About 80 calories are needed to melt one gram of ice or snow. In the northern half of the United States, between 5 and 10 percent of the incoming solar energy is used for melting during the spring season. Another example is forest fires, which give off about 850 calories per square centimeter on a single day, about 40 percent more energy than is typically available during a July day from the sun. Energy consumed by the friction between the wind and trees, mountains, and other land surfaces is usually not too important in balancing energy sources and sinks. Frictional effects are more significant as controlling factors in heat conduction in the air, wind flow, and pollution dispersion. On a global basis, less than 2 percent of the energy from the sun is used in the production of heat by friction.

The Big Picture

Up to now we have talked about the individual pieces of the weather-energy puzzle. Now we'll piece them together as they actually occur in the atmosphere to examine their impact on our daily weather. To better understand the relationship between weather conditions and energy processes at any instant of the day or night, we start by taking a look at conditions over an entire typical year, averaged over the entire world. We'll do this by forgetting about calories and referring only to relative energy amounts. The illustration on page 49 may be helpful in understanding the overall energy picture.

Since the sun is the source of all our weather energy, we start by dividing the sun's energy input (100 percent of the energy available to the earth for our typical year) into two parts: One part (54 percent) of the solar energy encounters skies that are cloudy; the second part (46 percent) strikes the earth's surface directly in cloud-free conditions. Some days are naturally neither clear nor totally cloudy and represent a composite of the two parts. Nonetheless, we'll proceed with the simple two-part picture, as it will provide a clearer understanding of the weather-energy picture without unnecessary complications.

The atmosphere is a heat reservoir of its own and, like the sun, radiates some of its energy to the earth's surface. The infrared energy received by the earth from the atmosphere amounts to nearly twice (83 relative energy units) what is received directly from the sun; this energy input is spread over each hour of each day, whereas the sun's inputs only come to us during the daylight hours. About 7 percent of the atmosphere's energy comes from solar energy that is absorbed by the air before it reaches the ground. We'll see later how the remainder is received.

A total of 115 energy units are received at the earth's surface in the form of solar (32) and infrared (83) radiation. Of the 32 solar units, 10 percent (3) are reflected and escape through the atmosphere to be lost in outer space; without this reflected light-energy, our astronauts would not be able to "see" earth without the use of special instruments. Because of the energy it receives, the earth also is a heat source, and like all "warm bodies," it

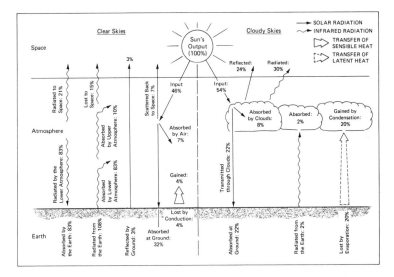

Distribution of energy between sun, earth, space, and the atmosphere

also radiates energy; at its world-average temperature of 50°F (10°C), the earth's surface radiates 108 units upward to air and space— 25 more than it received from the air. Most of the 108 units of radiated heat (83, to be exact) are absorbed in the lowest portions of the atmosphere. Another 10 units are absorbed higher up in the atmosphere, while the remaining 15 make their way out to space.

The ability of the atmosphere to absorb more than 75 percent of the infrared energy given off at the ground surface is due to the presence of water vapor, carbon dioxide, and other gases in the air. This effect has come to be called the *greenhouse effect,* although it is totally different from the processes that keep a greenhouse warm. There the warming effect is provided by the glass roof and walls, which let solar energy in but shut off heat loss by conduction in the air. This feature makes the daytime air in the greenhouse very hot, while at night the heat is lost by radiation. However, the heat accumulated during the day is so great that the nighttime losses cannot lower temperatures to damaging levels.

The greenhouse effect in the atmosphere is the center of the CO_2 controversy associated with the increased use of coal and other fossil fuels and the deforestation of much of South America and other forested regions. The use of coal to provide energy for new power plants will release increased amounts of CO_2 into the atmosphere. Trees, on the other hand, use CO_2 to produce food during photosynthesis. Together, the rapid depletion of large tracts of forested land and the world's appetite for more electricity will increase the CO_2 content of the atmosphere. This in turn will cause more of the outgoing infrared energy to be absorbed in the lower atmosphere, with the result that more energy will be reradiated back to earth. The end result could be an increase in the average temperature of the earth's surface. The increase might be only 1°F or less, but the impact on agriculture, weather, sea levels, and the ecosystem could possibly be significant.

Going through the numbers (again see illustration on page 49), we see that 4 energy units are unaccounted for at the earth's surface; more are coming in than we have yet to assign. Were these to be retained in the ground, there would be a continuing change in ground temperatures from year to year. But history shows this is not the case; ground temperatures are steady from one year to the next. This tells us that the remaining 4 units are given from ground to air by the process of conduction.

When you add up the energy units in the illustration, you will note that an equal amount of energy comes into and then leaves each of three zones in the earth-atmosphere energy system. These three zones are (1) the ground-air boundary, (2) the air-space boundary, and (3) the atmosphere. If these energy units did not balance, then either the earth or the atmosphere would experience a year-to-year temperature change and by definition we would be undergoing a change in climate. For climate to change there must be a steady change in the energy balance, although the amount or location of the energy gain (or loss) would be virtually undetectable unless it was averaged over a period of at least several decades.

Like the clear-sky portion of the energy pie, the cloudy portion is also in balance. Clouds reflect, absorb, transmit, and reradiate the energy they receive, thereby complicating the energy transfer processes. Clouds also are the source of precipitation, and consequently (1) they receive energy as water vapor condenses to form cloud droplets, while (2) the ground beneath them loses energy in the process of evaporating the moisture that has fallen from the clouds.

Up to this point, most of the discussion has been aimed toward understanding local factors. The sketches on page 52 carry this further by providing illustrations of the energy distribution and local weather conditions for three cases: clear-sky, midday conditions; clear-sky, nighttime conditions; and cloudy, nighttime conditions.

On any given day or night, the distribution of energy can vary from the average picture. As the energy distribution changes, so do the air and ground temperatures, local wind patterns, the chance for precipitation, and so forth. As mentioned at the outset, local weather conditions are the result of two factors: (1) the advection or transport of weather from upwind regions, and (2) the local distribution of the sun's energy. Either one can dominate at any given time, or your actual weather may be a combination of the two. The harder the synoptic-scale winds blow (that is, not winds that are of local origin, such as a sea breeze), the more important the regional weather pattern will be. And when winds are light, local conditions dominate. To forecast local weather successfully you must be able to understand the importance of each set of factors; know which dominates at the moment and will dominate in the future; and then be able to assess correctly how either local or regional factors will behave to control local conditions.

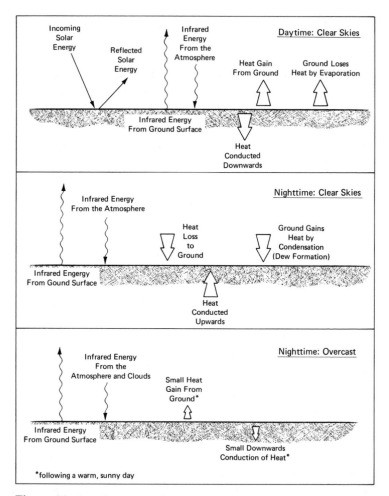

The partitioning of energy at the earth/atmosphere interface: three examples

Chapter Three

Cloud Watching and Weathercasting

O' look at the sky. A hundred clouds moving,
Gathered by puffs of wind on high,
Join their unknown forms. . . .

Then we think to see, in the swept sky,
A crocodile hang with a broad striped back,
And three rows of sharp teeth;
Beneath his lead belly floats an evening ray,
A hundred silver clouds shine below his black brow
Like golden scales.

Then a palace rises, the air trembles and all flee:
The frightful edifice of clouds destroyed
Falling in crowded ruins.

—VICTOR HUGO
Sunset I, *Autumn Leaves*

Poets have long lyricized the clouds, together with the anonymous authors of weather folklore that have given us numerous proverbs and sayings relating clouds to weather and weather to clouds. The fact of the matter is that clouds *are* weather, along with the winds, temperature, and precipitation. Yet clouds are much more than aerial reservoirs and sunshades. They are objects of beauty; ever-changing natural art forms that display endless variations in shape, texture, and hue.

53

Clouds are especially important to the outdoorsman for at least two reasons: First, they vividly depict the status and meaning of current weather conditions; and second, they provide telltale and reliable signs for predicting weather changes.

Before you can accurately predict tomorrow's weather, you must understand today's. Clouds can help you understand current weather conditions better than any other single weather element at your disposal. Watching their movement enables you to estimate the speed and direction of the wind at different altitudes. Observing present cloud forms (and recalling their predecessors) lets you judge the type, extent, and probability of precipitation. And by noting the cloud cover, thickness, and ceiling, you can reliably estimate maximum and minimum temperatures as well as the likelihood of fog and frost.

What Are Clouds?

Clouds are nothing more than a visible mass of water droplets or ice particles. These particles are usually so small that they float on the gentle wind currents in the air. The larger droplets fall out of the cloud. If they reach the ground, we have precipitation. Sometimes the air beneath the cloud is very dry or the droplets are not particularly big; then the droplets evaporate before reaching the ground producing a phenomenon called *virga.*

When the air is cooled sufficiently, the invisible water vapor it contains condenses onto tiny ice, salt, or other small particles to form liquid or solid water droplets. This same process occurs in your kitchen when the warm, moist air from the spout of a boiling teapot is cooled by the relatively cool air in the room. Clouds form in a similar way in the atmosphere, except that the cooling takes place in a different way. Clouds form when air ascends and cools as it expands because of the drop in pressure. This vertical lifting can result from air being forced up and over mountains or being carried aloft by convection currents on a hot day. Lifting also occurs as air is pushed up the gentle slope of a warm

54

front or the steeper slope of a cold front. On other occasions the cooling occurs as air moves horizontally from a warmer to a colder area (as from land to water in summer).

The shape and makeup of clouds reflect the way in which they are formed. Clouds created by gentle and widespread ascending currents generally have a thin, sheetlike appearance and are referred to as *stratiform* clouds. Towering clouds resembling cotton candy or cauliflower are formed by the more forceful updrafts of convection currents, mountainous terrain, or intense cold fronts; these are the *cumuliform* clouds. Low-level clouds consist of water droplets, while their highest neighbors—*cirriform* clouds—are aggregations of ice particles.

In some ways the state of the sky is like the state of the sea in that both are made up of a limited number of features (clouds or waves) that are easily recognized in isolation but become difficult and cumbersome to describe in combination. It is not too surprising, then, that more than 2,000 years elapsed from the early Greek writings about clouds by Thales, Anaxagoras, and Aristotle until the first comprehensive and accepted cloud classification scheme was devised in 1803 in England by Luke Howard. Another 128 years passed before the first truly international cloud atlas was published in 1930. And it was only in 1956 that the current classification came into being.

Although nearly a hundred cloud variations have been identified and named, they each belong to only one of ten basic cloud categories (or genera). This means that the clouds in each category can take on many appearances. Of course, at any time several cloud types may also appear in combination, and low-level clouds may obscure higher ones. But by properly identifying the cloud category, you will know some things about the cloud's height and composition, and the likelihood of precipitation (if any). And you can do this not only for the clouds you see in the sky, but also for those you will learn to expect over the next one or two days. In short, clouds are very good tools for making reliable, short-range precipitation forecasts.

Cloud Types

The ten basic cloud genera fall into four groups. Three of these groups are based on altitude: high, middle, or low. Clouds in these groups are usually quite flat. The fourth group consists of "towering" clouds that are easily recognized by their extensive vertical development and their rapid growth.

High-altitude clouds are of three types: *cirrus, cirrostratus,* and *cirrocumulus.* In temperate latitudes they are found at very high altitudes, ranging from 3 to 8 miles, although they are extremely thin (less than 100 feet thick). Because of their great height, all three types consist almost entirely of ice crystals. Cirrus clouds are recognized by their white, delicate filaments and silky sheen. They may consist of tufted patches or narrow bands. Cirrostratus differs in that it covers most or all of the sky in a whitish, transparent veil. Both cirrus and cirrostratus can often be distinguished by the halo they create around the sun or moon. Cirrocumulus is easily recognized by its overall patchy or sheetlike appearance and its speckled or scaly texture (caused by the small cloud globules that group together to form this cloud type).

Cirrus can be either a sign of fair weather *or* the first herald of an approaching storm system. Very high cirrus elements that are thin and wispy are a good indicator of fair weather, particularly if they are *not* followed by cirrostratus. Other cirrus types signal impending bad weather: Cirrus arranged in parallel bands or stripes indicate the possibility of very poor weather ahead; mares' tails are another first sign of potentially poor (though less severe) weather; and so-called windy cirrus resemble mares' tails but are lowe. and more diffuse, and serve as a preliminary warning for stormy and very windy weather.

As a general rule, the higher cirrus clouds foretell fair weather, while the lower ones are the first sign of stormy weather. Even so, these "bad weather" cirrus should only be taken as a warning to be on the lookout for additional weather signs. But when they are followed within a few hours by an expansive sheet of cirrostratus, the chances for rain within 24 hours rise to about 80 percent. The reason is that the cirrus and cirrostratus mark the first, upper-level reaches of a warm front. As the lower portions of the front follow, middle- and low-level clouds will appear with their precipitation.

Middle-altitude clouds are of two types: *altostratus* and *altocumulus*. Altostratus cover the sky with a blue-gray veil, giving an effect commonly called a "watery sky." Although light rain or snow can fall from them, the presence of altostratus is more important as a forecasting aid. When altostratus appear several hours after cirrus and cirrostratus, the chances for steady precipitation in the next six to twelve hours jump to 90 percent. In a contrary way, altocumulus should not be interpreted as a forewarning of precipitation from weather fronts. The garden-variety altocumulus is distinguished by the way its moderate-to-large puffs are arranged in groups or lines. When shaped like large cotton balls, they are often referred to as "sheep backs"; at other times, the altocumulus cloud elements take on larger, raggedy forms. It is not uncommon to see patches of altocumulus intermingled with a broad layer of altostratus.

The three basic forms of low-level clouds are: *stratocumulus, stratus,* and *nimbostratus.* All three are capable of producing precipitation, but the amounts and duration are greatest with nimbostratus. For this reason, nimbostratus clouds are commonly called rain clouds. They are low and thick and often have a deep gray coloring, and their undersides are frequently marked by ragged patches (called *scud*) that detach themselves from the main cloud layer. Stratocumulus is seen most often in winter. Its large, soft globular cloud elements combine in groups or waves to cover large portions of the sky. Stratus is a soft, foglike cloud that frequently produces a drizzle, but never rain. It often begins as fog and is transformed into stratus when the fog lifts or its lower layers dissipate.

Perhaps the most beautiful, and often the most dangerous, clouds belong to the towering group: *cumulus* and *cumulonimbus.* Occurring almost exclusively on warm and sunny days or along the imaginary line that marks a cold front, cumulus clouds are the easiest type to recognize. Their undersides are quite flat, grayish, and low (1,500 to 5,000 feet), while their cauliflower sides can tower a mile high, glistening white in the high midday sun. As cumulus are formed by convection currents, they are most frequently found inland and along the coast, but usually not far out at sea. They form in the morning and then grow in number and size until at times they dot the entire sky by midafternoon. They then die rapidly with the late-afternoon sun. At sea, convection is less intense and often lacking altogether, and cumulus are produced instead by the rising air of cold fronts or the action of the evening offshore land breeze.

The common cumulus cloud produces little if any rain other than an occasional short-lived shower. One variation, called *cumulus congestus,* is worth noting not only because it can produce nominal rainfalls but because it is often the predecessor of a thunderstorm. The cumulonimbus is indeed a dangerous beast, unleashing its fury in many ways: high winds, heavy rains, hail, and lightning. Near its mile-high base, air rises into the storm cloud at speeds of around 2 mph; but at 25,000 feet it is not unusual for the air to rise at 60 mph. As precipitation falls out of the cloud, cold air is dragged along, causing ground-level temperatures to drop by as much as 20° to 30°F in a matter of minutes. Surface winds may gust to speeds of 50 mph and greater. But with the downdraft, the storm begins to die and its ice-laden, anvil-shaped top starts to break up. In its death throes, however, the cumulonimbus is often at its worst, unleashing a broadside of lightning, hail, heavy winds, rain, and possibly even a tornado. The way to avoid these dangers is clearly to recognize the cumulus and cumulus congestus cloud forms that precede the cumulonimbus, and then to move out of its way or seek refuge.

TABLE 7. TERMINOLOGY DESCRIBING THE STATE OF THE SKY

Term	Cloud Cover
Clear	Less than $1/_{10}$ (10%)
Scattered, or partly cloudy	$1/_{10}$ to $5/_{10}$
Broken, or cloudy	$6/_{10}$ to $9/_{10}$
Overcast	More than $9/_{10}$

TABLE 8. SOME CHARACTERISTIC FEATURES OF SIX COMMON* PRECIPITATION-PRODUCING CLOUD TYPES

	Alto-stratus	Nimbo-stratus	Strato-cumulus	Stratus	Cumulus	Cumulo-nimbus
Precipitation						
Rain	possible	usual	possible	—	possible	usual
Drizzle	—	—	—	possible	—	—
Snow	possible	possible	possible	—	—	possible
Hail	—	—	—	—	—	possible
Other Features						
Halo	—	—	—	possible	—	possible
Rainbow	—	—	—	—	possible	possible
Thunder and lightning	—	—	—	—	—	possible

* Four cloud types that do not produce precipitation have not been included: cirrus, cirrostratus, cirrocumulus, and altocumulus.

TABLE 9. SUMMARY OF BASIC CLOUD TYPES AND THEIR FEATURES

CIRRUS (feathery clouds)—high, thin, and wispy clouds. Usually form above 25,000 feet where temperatures are well below freezing; they are composed entirely of ice crystals. They do not produce precipitation, but can often be recognized by the partial halo they form around the sun or moon. Cirrus arranged in bands and those that resemble mares' tails often signal approaching storm systems, especially when followed by cirrostratus or altostratus.

CIRROSTRATUS (halo-producing)—usually form between 20,000 and 25,000 feet. Appear as thin whitish sheets resembling veils. Because of their broad expanse and ice-crystal composition, cirrostratus usually form halos around the sun and moon. They produce no precipitation, but foretell storms when they occur in a procession between cirrus and altostratus.

CIRROCUMULUS (mackerels' scales)—also form between 20,000 and 25,000 feet and consist mostly of ice crystals. Thicker than cirrostratus, they do not form halos nor do they produce precipitation. They are a transient cloud formation, often forming either from cirrus or cirrostratus and then quickly returning to these other forms. Not an indicator of weather changes, cirrocumulus usually accompany fair weather.

ALTOSTRATUS (watery sky)—composed of liquid water droplets; they have their bases around 15,000 feet and are up to 3,000 feet thick. Altostratus look like dense veils, or sheets, of blue-gray that often wash out the sun. Light rain or snow is possible, but not common. When preceded by cirrus and cirrostratus, altostratus indicate a very high chance for rain-bearing nimbostratus in the next 6 to 12 hours.

ALTOCUMULUS (sheep backs)—with bases above 10,000 feet, these large globular cloud masses are composed of water droplets and are usually arranged in groups or lines. They appear white or pale gray and the edges of the clouds are often indistinct. Altocumulus do not produce precipitation. When followed by low-level towering cumulus clouds, they are a sign of impending stormy, showery weather.

STRATUS (layers or sheets)—a low, uniform layer that resembles fog, but does not rest on the ground. Cloud bases are very low, varying from heights of a few tens of feet to 6,500 feet. Some precipitation may result from stratus, but only in the form of a light drizzle.

STRATOCUMULUS (long, flat layers)—irregular masses of large, puffy clouds that are spread out in broad, undulating layers. They often form at day's end from midday cumulus clouds, and are an indication of a clear and cool evening. Some light rain or snow is possible, but is uncommon and short-lived. Stratocumulus are most common in winter, when their bases may be as low as 1,000 feet high.

NIMBOSTRATUS (rain cloud)—low, uniform, and thick clouds. These dark gray clouds are the most common rain (or snow) producers. Precipitation from nimbostratus is steady and prolonged. Most commonly found along weather fronts, these clouds may be a mile thick with bases that rest close to the ground (500 to 6,000 feet).

CUMULUS (woolpack)—easily recognized by their flat, gray bottoms and white, puffy sides. Their bases vary from 2,000 to 14,000 feet high. A brief, light sprinkle is possible from cumulus, but not common. The lower and thicker the cumulus are, the higher the chance that they will transform into rain-yielding cumulus congestus or cumulonimbus. When the sky is dotted with rows of small, puffy cumulus, it is a sure sign of fair and dry weather that day.

CUMULUS CONGESTUS (cauliflower)—not a basic cloud type, but a member of the cumulus family. An important cloud to recognize, not only because of the rain showers it can produce but also as a forewarner of more severe weather. When mushrooming cumulus congestus are seen in mid- to late morning, there is a good chance of afternoon thunderheads.

CUMULONIMBUS (thunderhead)—an intense storm cloud that may be several miles across with its top reaching from 3 to 10 miles high. It is marked by an anvil-shaped top that is made up of ice crystals. Cumulonimbus are usually formed along intense cold fronts, on the upwind side of mountains, or on sultry summer days. They are accompanied by thunder and lightning. A single cloud can produce torrents of rain and hail for an hour or more, with the record one-minute downpour standing at 1¼ inches.

MAMMATUS (pouches)—a cloud feature that can be found on any of several cloud types. Indicates a very unstable cloud formation, especially with cumuliform clouds. Particular caution should be exercised when mammatus is seen as a feature of thunderheads, for the storm is apt to have high winds and hail.

Chapter Four

Large-scale
Weather

The emphasis of the preceding chapters has been on weather basics: temperature, pressure, winds, energy, and so forth. On many occasions, the weather you experience is the direct result of local effects due only to the supply of solar energy at your location, the properties of the ground surface (for example, smooth or rough, wet or dry, flat or sloped), and the nature of the air around you (hot or cold, dry or humid, clear or cloudy). But as often as not, these local effects will not shape the nature of the weather where you may be; instead, your weather will be the product of atmospheric events that took place days before and thousands of miles away. A cold spell in Florida may be the result of air imported from northern Canada; a heat wave in New England may be air that was first warmed over the Gulf of Mexico; and a squall line with violent thunderstorms in the upper Midwest may have been spawned by the collision of Gulf and Arctic air masses. Clearly in all three cases the local weather was the end product of large-scale atmospheric features and conditions. Nearly all of the weather features shown on the weather map in the newspaper or on television are large-scale weather; these include fronts, pressure systems (highs and lows), and satellite "photos" of cloud patterns. Only when the TV weathercaster shows you, say, the spread of temperatures throughout the local area are you seeing the results of some of the small-scale effects discussed earlier; these effects are the main focus of Part Two. For now, however, the weather effects of localized terrain features will be ignored as the nature and impact of large-scale weather are explored.

When planning a trip to an unfamiliar location, a road map and a geographical atlas are invaluable aids. They tell you what routes are available to you, how long the trip will take, what cities are along the way, and what worthwhile stops there may be. In essence, the map and atlas let you anticipate, plan, or predict the character of your trip. Think of the weather map as an atmospheric map-atlas. Instead of following a superhighway from your home to your destination, weather follows a path from some location *to you*. The weather map can tell you what weather conditions are like across the country, how they are apt to change, and where tomorrow's weather in your area is likely to originate and what it is apt to be. The weather map can also tell you more. It can tell you, for example, when local conditions will dominate the weather scene and when they won't. With this knowledge you will be better prepared to forecast weather at the location that is most critical to you—at your campsite, along your charted course, or in your hammock.

Air Masses and Pressure Systems

The most fundamental member of the family of large-scale weather systems is the air mass. As it develops and matures it can create a frontal storm system or present itself as a fair-weather, high-pressure system.

Large expanses of air with characteristic pressure (and therefore wind) patterns fall into one of two basic categories: *high-* or *low-pressure systems.* High-pressure systems often breed air masses because "highs" tend to have the weak wind flows that keep air in an area long enough to form a uniform air mass. A high is usually several hundred miles in diameter and has high pressure at its center and lower pressure toward the edge. This pressure distribution causes a clockwise or anticyclonic flow of wind around the core of the high, and therefore highs are also called *anticyclones*. Winds are generally light and variable at the center and are often quite strong at the edge of the high. In addition to moving in an expanding circular pattern, the air in a high moves downward at the center. This usually results in clear skies and a small probability of precipitation.

A "low" or *cyclone* is the counterpart of a high. Air pressure is relatively low at the center and increases toward the outside of the low. Wind circulation about the center is counterclockwise or cyclonic, with the strongest winds often found at the center. A low often has strong upward motion at its center and consequently usually means cloudy weather, sometimes producing heavy amounts of precipitation. In contrast to highs, lows do not have uniform characteristics over their widths (sometimes 500 to 1,000 miles across). Instead they are marked by boundaries called *fronts* where wind, pressure, humidity, temperature, and cloud features change dramatically in the space of a few miles.

High- and low-pressure areas are not always large, expansive systems with gently curved isobars. Sometimes they occur on a much smaller scale and appear as tongue-shaped masses when their isobars are plotted on a weather map. In the case where they are characterized by relatively high pressure at the center, they are called *ridges* or *wedges;* their low-pressure counterparts are called *troughs.* Neither ridges nor troughs have isobars that make closed patterns. Ridges often produce strong winds though little or no precipitation. Troughs, on the other hand, may yield some rain or snow, though it is usually light and short-lived.

Both highs and lows generally move from west to east across the United States in an alternating sequence. Sometimes they may move at a speed of only 1 or 2 mph (or not at all), while at other times they zip along at 20 to 30 mph.

The term "air mass" means nothing more than a broad envelope of air that has nearly uniform temperatures and humidities (to be more precise, specific humidity). Were we dealing with people rather than air, we might refer instead to an ethnic group. If we selected Italians as an example, we might characterize the group by its geographical origin, its language, and its members' characteristic physical features. When we deal with air masses, we label them according to:

- Temperature of the air relative to the ground
- Moisture (humid or dry)
- Geographical origin (for example, Arctic)

An air mass is normally very large; on the small end it might cover an area equal in size to the state of Texas, while the larger ones can encompass an area ranging from the Rocky Mountains eastward to the Atlantic Ocean and from Canada southward to the Gulf of Mexico. Not only are they very broad, but they are deep as well. The more shallow air masses extend from the ground to several thousand feet, while the larger, more mature ones reach several miles.

How does an air mass form? When a large expanse of land or water is calm for a period of several days, the layers of air in contact with the earth's surface take on temperatures and humidities similar to those of the surface. If the surface is frozen tundra, the overlying air will be cold and dry. Ocean air, on the other hand, may become warm and moist. The influence of the surface is felt through deeper layers the longer the air sits over the same type of surface. As a wind picks up, the mass of air may move to another region, or it may move in a large circular pattern, in which case the air mass would continue to mature, becoming deeper, broader, and more like the underlying surface. But in all cases it is essential that the surface area be sufficiently large to ensure that the air has a lengthy residence time over it and sufficiently uniform to ensure that the air also becomes uniform and hence distinct. If the surface is not uniform, then any movement of the air from one surface type to another will result in its modification so that the air no longer has a unique identity.

Let's assume an air mass has moved from its source region into a mountainous area and you are about to go from the base to the top of the mountain. What can you expect? To answer this question requires some thought about the vertical structure of the air mass. If the air has been cooled from below by contact with or radiation from a cold surface, then the air temperature may actually increase with height. Conversely, a warm surface will have decreasing temperatures aloft. Humidity, on the other hand, will usually decrease with height within a single air mass. So, while air masses are uniform horizontally, they are not uniform vertically. Also, since an air mass may range from 1 to 3 miles (1.6 to 5 km) in thickness, it is possible that we may pop out of one air mass and into another if our hypothetical mountain is tall enough.

Where do most American air masses form? Of the six air-mass source regions, only one is actually located within the boundaries of the contiguous forty-eight states; the other five are found over the Gulf of Mexico, the Atlantic and Pacific oceans, and the Canada–Alaska–Arctic Ocean region. Two form over land: The tropical continental air mass of the southwestern states only forms in summer, while the polar continental air masses north of the border form throughout the year. Of the four maritime air masses, two form in the Pacific, one in the North Atlantic, and one in the Gulf–South Atlantic area. With the exception of the summertime maritime Pacific air mass, the source regions are not continuously inhabited by their air-mass namesakes. Rather, the air masses go through cycles. First an air mass will form and then it will migrate to neighboring or more distant areas.

How do air masses change as they move? Air masses can change in one of three ways. First, they can be modified from below in much the same way as they originally formed: by moving over a large area with different surface features. Second, a moist air mass may move up and over a mountain range. As it is lifted it may cool to the point where condensation and precipitation occur.

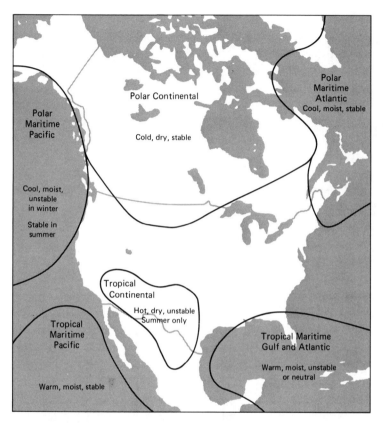

Source regions for North American air masses

The air becomes drier through the loss of moisture and warmer because of the energy that is released in the condensation of the humid air. Third, dissimilar air masses may converge to form weather fronts, drastically changing the air-mass features along the frontal zones. The number of possible ways in which an air mass can actually be modified is enormous because of the various combinations that can exist among the three basic modifying mechanisms.

TABLE 10. CLASSIFICATION OF AIR MASSES

Source by—		Local Source Region	Corresponding Air-Mass Name	Typical Temperature Characteristics*
Latitude	Nature			
Polar	Continental	Alaska, Canada, and Arctic	Polar Canadian or continental	Winter: warmer or same Summer: colder
		Modified in southern and central United States	Transitional polar continental	Winter: warmer or colder Summer: colder
	Maritime	Colder portions of North Pacific Ocean	Polar Pacific	Winter: colder Summer: warmer or same
		Modified in central and western United States	Transitional polar Pacific	Winter: warmer Summer: colder
		Modified over warm portions of Pacific	Transitional polar Pacific	Winter: colder or same Summer: colder or same
		Colder portions of North Atlantic Ocean	Polar Atlantic	Winter: colder Summer and spring: warmer
		Modified over warm portions of Atlantic	Transitional polar Atlantic	Winter, spring, and summer: colder

Tropical	Maritime	Gulf of Mexico, Caribbean Sea, Sargasso Sea, and Middle Atlantic (also southern United States in summer)	Tropical Atlantic	Winter: colder or same Summer: colder
		Modified over northern U.S. or North Atlantic	Transitional tropical Atlantic	Winter and summer: warmer
		Northern part Pacific trade-wind belt	Tropical Pacific (usually not found in summer)	Winter: colder or same
		Modified in U.S. or over North Pacific	Transitional tropical Pacific	Winter: warmer

*Temperature of the air mass in comparison with the surface over which it is traveling

Fronts and Cyclones (Storms)

We frequently hear about fronts—cold fronts, warm fronts, and sometimes stationary and occluded fronts. Fronts themselves are not a type of weather, nor do they always bring bad (or good) weather with them. However, they often are a cause of bad weather and although there are other causes as well, it remains that if we are to understand and predict weather, we need to understand and predict the movement of fronts.

First a few basic facts about fronts. They are not walls, nor are they invisible dividing lines. Rather they are zones that mark a change or transition between air masses with differing properties. Fronts are typically anywhere from a few miles to 60 miles wide (100 km) and extend from the earth's surface to the top of the shallower of the two air masses. The frontal zone does not extend upward at a right angle to the earth's surface, but slopes away from the warm side of the frontal zone. As we'll see later, this means that the ground-level portion of a cold front reaches any given location before the upper levels of the front, while the upper reaches of a warm front (made visible by clouds) arrive before the surface portion. Fronts vary in length from a few tens of miles to 500 miles or more in some cases. They can travel at up to 40 mph or they can be immobile.

Fronts usually form at the boundary of two dissimilar air masses, as illustrated on page 72. Cold, dry, and stable air is characteristic of the air mass found in the northern portions, and warm, moist, and unstable air lies southward. At some point along the boundary the winds may not blow parallel to the boundary, but rather blow toward opposing winds in the adjacent air mass. A bulge develops between the two air masses, and finally a wavelike pattern evolves along the boundary. The pattern and the forces that produce it are very similar to those of ocean waves. At the water-air boundary, waves form that resemble the frontal pattern between the colder, denser air to the north and the warmer and moister (and consequently lighter) air to the south.

Depending on the nature of the air masses and the portion of the boundary zone, several different types of fronts can be identified: stationary, warm, cold, and occluded.

As its name implies, a *stationary front* simply marks an immobile boundary between two air masses. The front may have been a

---

TABLE 11. AIR MASS PROPERTIES

TYPICAL PROPERTIES OF COLD AIR MASSES

Property	Continental	Maritime
Humidity	Fairly constant	Increasing
Clouds	Scattered cumulus, occasional cumulonimbus	Plentiful and heavier cumulus and cumulonimbus
Precipitation	Light showers, mostly in afternoon	Heavy, squally showers, mostly in early morning
Cloudiness	Pronounced diurnal variation, maximum in afternoon	Slight diurnal variation, mostly in early morning
Sky	Considerable bright intervals	Variable between bright and threatening
Cloud base	Considerable; seldom below 2,000 feet	Moderate, but seldom below 1,000 feet
Visibility	Variable, mostly good, except for dust and smoke	Excellent, between showers

TYPICAL PROPERTIES OF WARM AIR MASSES

Property	Continental	Maritime
Humidity	Low, fairly constant	High, increasing
Clouds	Very few, high	Hazy, stratus type or fog
Precipitation	Little if any	Steady, light
Cloudiness	Clear or a few scattered	Variable to overcast
Visibility	Good	Poor, hazy

SOURCE: Adapted from C. G. Halpine and H. H. Taylor, *A Mariner's Meteorology* (Princeton, N.J.: D. Van Nostrand Co., Inc., 1956), p. 141.

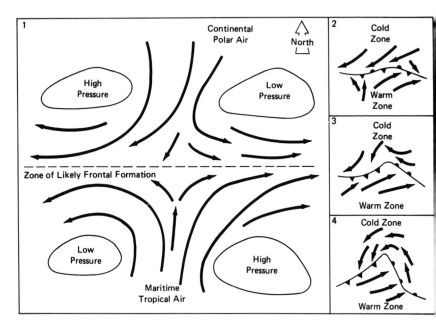

Four phases in the life of a frontal system. *Fronts frequently form between converging polar and tropical air masses where the pressure distribution and wind patterns resemble those illustrated here.*

warm front in the past if it moved in such a way that warmer air replaced cooler air at the surface, or it might develop into a cold front if it starts to move the opposite way. But as a stationary front it does not move. The winds to either side normally blow in opposite directions. As such the stationary front commonly spawns the development of cold and warm fronts as described earlier. Stationary fronts have no foul weather or clouds associated with them.

Along a *cold front*, colder or cooler air displaces warmer air as the front moves. A place where a cold front has passed would have first experienced relatively warm air that was subsequently replaced with colder air. The temperature change may be only a few degrees or as much as 15° to 20°F (8° to 11°C); as the temperature drops further, the chance of foul weather increases. Cold fronts are usually accompanied by cumulus-type clouds and heavy, showery precipitation.

Warm fronts are characterized by the replacement of cold air by warm air. Warm fronts are broader and less intense than their cold counterparts. The frontal surface slopes gently upward to the east-northeast with a vertical rise of one mile for every 150 miles along the ground. A fully developed warm front may stretch nearly 1,000 miles from its uppermost, easterly portions 6 miles high to its ground-level sector to the west. While New York sees the associated high-altitude cirrus clouds that herald its arrival, Memphis receives the steady rains from the low-level nimbostratus cloud deck.

Weather Elements	Cold Front		Warm Front	
	Before	After	Before	After
Wind direction	Southwest	Northwest	South	Southwest
Wind speed	Moderate	High	Low-moderate	Moderate
Temperature	Warm	Cold	Cool	Warm
Clouds	Cumulus	Clear	Stratus	Cirrocumulus
Precipitation	Yes—heavy	No	Yes—moderate	Yes—showery
Humidity	High	Low	Moderate	High
Pressure	Low	High	High	Low

TABLE 12. CHARACTERISTIC WEATHER CHANGES ASSOCIATED WITH THE PASSAGE OF COLD AND WARM FRONTS

Cold fronts often travel faster than warm fronts. When a storm system (also called a cyclone) has both a cold front and a warm front, the cold front is always initially to the west of the warm front. And as we'll see shortly, the two fronts are usually joined

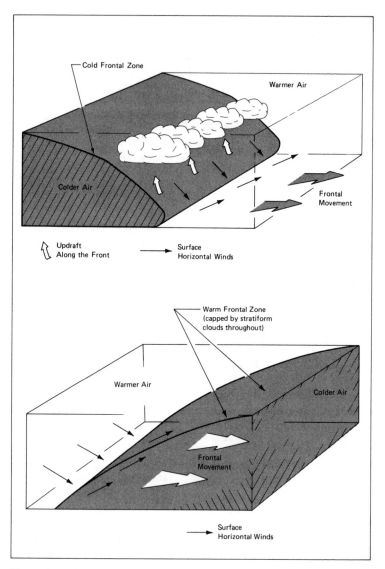

Three-dimensional sketches of the structure of a cold front (top) and a warm front (bottom)

wishbone-fashion at their northerly ends. When the cold overtakes the warm, the result is an *occluded front* that separates the cool air ahead of the warm front, the cool air behind the cold front, and the warm air between the two fronts. Two types of occlusions can result: a *warm-front occlusion* and a *cold-front occlusion.* The type actually found depends on which of the two cool (or polar) air masses is colder. If the cool air behind the cold front is colder and denser, then a cold-front occlusion results. If the air ahead of the warm front is colder than that behind the cold front, a warm-front occlusion is the result. With both occlusions the weather pattern is virtually identical, resembling a warm front ahead of the front and a cold front as it passes, with no clearing zone between. Occlusions usually occur in the late stages of a frontal system; the intensity of the system peaks as the occlusion process starts and begins to deteriorate as the occlusion matures. (The temperature difference across an occluded frontal zone is between cold and cool air, which produces less intense weather than the cold-warm air difference across warm and cold fronts.)

Earlier we discussed the formation of frontal zones between dissimilar air masses. When fronts form in this way it simultaneously marks the development of a cyclonic disturbance—also called a cyclone, low, or storm system. The wavelike pattern that marks the boundary between the different air masses is the core of the storm system. Atmospheric pressure is lowest at the tip or crest

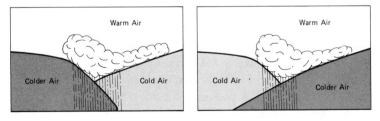

Cold-Front Occlusion   **Warm-Front Occlusion**

Cross-sectional views of occluded fronts

of the wave where the cold and warm fronts meet. As the storm matures, the fronts become more intense and better defined, atmospheric pressure drops at the center, and winds, clouds, and precipitation increase. Not all storms are alike. Some have a cold front but no warm front; some fronts occlude, while others may weaken and disappear. Often several storms and their frontal systems will be linked across the country like a tinsel garland hanging from the branches of a Christmas tree. With satellites now a routine tool for the weathercaster, storm systems can easily be identified and tracked day and night by the cloud patterns they invoke. They are also easily recognized by the wind, pressure, temperature, and precipitation patterns they make on a weather map. Although they are often unpredictable, there are a number of tracks or roadways that these storms take across the United States. Of eight major storm tracks, only two span the entire country: one from the Pacific Northwest through the upper Midwest to New England; the other from southern California through the lower Midwest and into the central eastern seaboard. Three of the remaining six originate on the eastern slope of the continental divide, one in western Canada, another in Kansas, and the last in the Gulf of Mexico.

The Weather Map and Some Simple Forecasting Tips

The weather map, or *synoptic chart,* is the main source for getting a general picture of weather patterns over a large area. It is usually readily available from newspapers and television, and, with the help of your imagination, even VHF weather broadcasts (see Appendix C). The map of weather conditions at ground level ("surface map") prepared by the National Weather Service shows how the various weather elements are distributed around the country. The map of ground-level weather shows:

Temperature
Dew Point
Pressure
Pressure change

Wind speed and direction
Cloud type, cover, and height
Current weather
Precipitation (rain, snow)

These elements are plotted on the map at the location of each observing station. On the same map, lines connecting locations with the same pressure readings are also shown. These isobars outline high- and low-pressure systems, as well as warm, cold, stationary, and occluded fronts. Fronts are often identified on the weather map by distinctive colors: a solid blue line for a cold front, red for a warm front, alternate blue and red for a stationary front, and purple for an occluded front. Regions with widespread precipitation are easily recognized by shaded areas on the weather map.

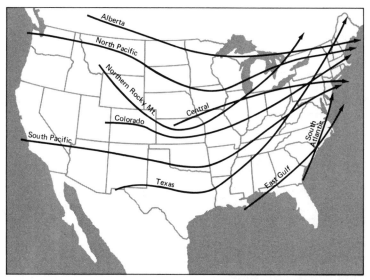

Major storm tracks across the United States. *All go from west to east, except for the "South Atlantic" track that goes northward along the East Coast; nearly all storm tracks converge on the northeastern states.*

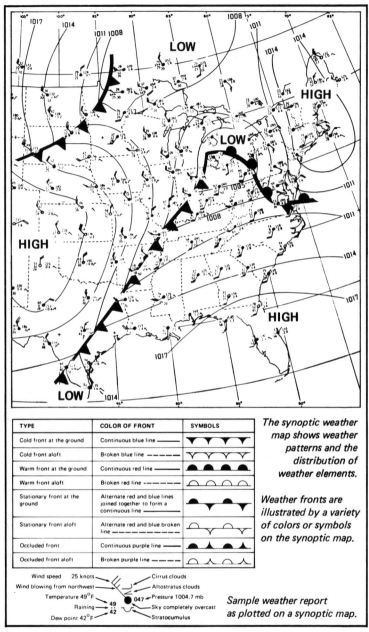

TYPE	COLOR OF FRONT	SYMBOLS
Cold front at the ground	Continuous blue line ———	▼▼▼▼
Cold front aloft	Broken blue line — — — —	▽▽▽▽
Warm front at the ground	Continuous red line ———	●●●
Warm front aloft	Broken red line - - - - -	◠◠◠
Stationary front at the ground	Alternate red and blue lines joined together to form a continuous line ———	●◠▼
Stationary front aloft	Alternate red and blue broken line — — — — — —	◠ ▽ ▽
Occluded front	Continuous purple line ———	●◠●
Occluded front aloft	Broken purple line — — —	◠▲◠▲

The synoptic weather map shows weather patterns and the distribution of weather elements.

Weather fronts are illustrated by a variety of colors or symbols on the synoptic map.

Wind speed   25 knots ——— Cirrus clouds
Wind blowing from northwest ——— Altostratus clouds
Temperature 49°F ——— Pressure 1004.7 mb
Raining ———
Dew point 42°F ——— Sky completely overcast
——— Stratocumulus

Sample weather report as plotted on a synoptic map.

Weather map and symbols

Not only does the weather map give you a snapshot view of current weather conditions, it can also help you forecast the weather for periods of up to twenty-four hours. First look at the map to find the location of the frontal systems and see whether there are reports of precipitation near them. Then see how "strong" the pressure difference is between the center of the low-pressure region and its edge. A large pressure difference is one sign of an intense storm system. Next, if you have an earlier weather map handy (no more than a day old), you can locate the previous positions of the same fronts and determine how fast they have moved. Moving them ahead at the same speed and direction will give you a good idea of what kind of weather your area could experience over the next day. Fronts usually move in a general west-to-east direction at speeds as fast as 40 mph, but a typical speed is around 20 mph in summer and 30 mph in winter.

This forecasting technique is called *extrapolation,* and it is based on the theory that the general features of today's weather scene will not change drastically but that they will move today as they did yesterday. Of course, extrapolation doesn't always work, although it can provide a decent enough forecast most of the time. If you try the extrapolation method and find you were wrong, don't be discouraged. You may be able to improve your forecast skill by understanding what produced your incorrect forecast. Something happened differently today than it did yesterday. Perhaps a ridge of high pressure blocked the storm's movement or changed its direction. Or maybe a dying storm system came to life again when it moved away from land over the ocean. Consult tomorrow's map and it will show you what happened. Then next time you may be able to see the change coming and include it in your forecast.

The extrapolation method will provide an accurate 12- to 24-hour forecast about three times in four. Simply assuming that tomorrow's weather will be the same as today's (persistence) is accurate about two times in three. For comparison, the computerized weathercast prepared by the National Weather Service is correct about five times in six.

The most difficult forecast to make is when weather patterns change. Often, the weathercaster suspects that one of two or three possibilities will cause a change in tomorrow's weather. To tell people how much confidence he has in his forecast, he will often assign a probability. For example, "Skies will be mostly cloudy tomorrow afternoon, with a 30 percent chance of rain toward evening." Based on the forecaster's experience and the information he has on hand, he expects that it would rain three times for every ten times he made that same forecast.

Barometric pressure is perhaps the single best indicator of forthcoming weather changes. Two types of pressure signs are particularly helpful in making short-range weathercasts: the change of pressure with time (say, three or six hours), called *pressure tendency,* and the change of pressure over distance, called *pressure gradient,* as described in Chapter 1.

To keep abreast of pressure changes, you will need a reliable barometer, with a reference pointer. About once a year it should be checked. A phone call to the National Weather Service will give you the current pressure, adjusted as it would read at sea level.

How fast pressure changes at your location is a sign that a pressure system is either moving or changing character. When pressure systems move, they may bring with them either clear weather and higher pressure or poor weather and lower pressure. But a weather system can remain stagnant and still cause a pressure change: A "deepening" situation is when the pressure drops steadily. This is a sign that a storm is in the making and that poorer weather is ahead. The opposite situation, when pressure rises, is called "filling," and means fairer weather is ahead.

Meteorologists and other weather-watchers have been studying the relationship between pressure tendency and weather changes for more than a hundred years. They have found that a fairly reliable short-term forecast can be made when a simple observation of wind direction is added to pressure and pressure tendency readings. No fancy wind-measuring equipment is necessary. You can read wind direction by checking which way a simple wind vane points or even just watching a flag or a puff of smoke. The Wind-pressure Forecast Chart (Table 13, pages 82 to 83) can then be used with some confidence.

Weathercasting based on wind and pressure information at one place can be done with some success. But in today's world we usually have access to additional information. For example, the weather map provides information on the pressure gradient, which is easily estimated by the distance between isobars. A tight gradient (isobars close together) or one that is tightening is a sign of foul weather. And if the gradient tightens within a low-pressure system, precipitation can be expected along with poor visibilities and low ceilings (heights of the cloud base).

When the pressure gradient is flat (isobars far apart), wind speeds are low and there is little or no chance of rain or snow. Skies are generally clear, except that fair-weather cumulus clouds may form on summer days. However, the ground and the lower layers of air cool rapidly at night, creating a temperature inversion. This is the weather type most likely to lead to episodes of heavy air pollution.

Pressure systems influence the weather in two ways. First, each has its own characteristic weather features. Second, they can affect the overall synoptic patterns by blocking other systems.

High-pressure systems are generally marked by the absence of precipitation. Winds are light and variable in the center, but they can be strong at the edge. Humidities are low and temperatures are below seasonal normals. Low-pressure systems, on the other hand, frequently carry precipitation with them. They are the home of the weather front. In low-pressure systems, winds are strong, humidity is high, and clouds abound.

Stagnating highs (those that plant themselves over a region for days at a time and scarcely move) frequently locate in the northeast United States in springtime. Their effects can be felt as far west

TABLE 13. WIND-PRESSURE FORECAST CHART

Wind Direction Change	Sea-Level Barometric Pressure (inches of mercury)	Pressure Tendency	Character of Weather Indicated
SW to NW	30.10 to 30.20	Steady	Fair with slight temperature changes for 1 to 2 days
SW to NW	30.10 to 30.20	Rising rapidly	Fair, followed within 2 days by rain
SW to NW	30.20 and above	Steady	Continued fair with no decided temperature change
SW to NW	30.20 and above	Falling slowly	Slowly rising temperature and fair for 2 days
S to SE	30.10 to 30.20	Falling slowly	Rain within 24 hours
S to SE	30.10 to 30.20	Falling rapidly	Wind increasing in force, with rain within 12 to 24 hours
SE to NE	30.10 to 30.20	Falling slowly	Rain in 12 to 18 hours
SE to NE	30.10 to 30.20	Falling rapidly	Increasing wind, and rain within 12 hours
E to NE	30.10 and above	Falling slowly	In summer with light winds, rain may not fall for several days; in winter, rain within 24 hours
E to NE	30.10 and above	Falling rapidly	In summer, rain probable within 12 to 24 hours; in winter, rain or snow with increasing winds will often set in when the barometer begins to fall and the wind sets in from the NE

		TABLE 13 *(continued)*		
Wind Direction Change	Sea-Level Barometric Pressure (inches of mercury)	Pressure Tendency	Character of Weather Indicated
SE to NE	30.10 and below	Falling slowly	Rain will continue 1 to 2 days
SE to NE	30.00 and below	Falling rapidly	Rain with high wind, followed within 36 hours by clearing and colder temperatures in winter
S to SW	30.00 and below	Rising rapidly	Clearing within a few hours; fair for several days
S to E	29.80 and below	Falling rapidly	Severe storm imminent, followed within 24 hours by clearing and colder temperatures in winter
E to N	29.80 and below	Falling rapidly	Severe northeast gale and heavy precipitation; in winter, heavy snow followed by a cold wave
Going to W	29.80 and below	Rising rapidly	Clearing and colder

as the Mississippi Valley. They act like a traffic signal, blocking the eastward movement of storm systems and causing poor weather over much of the eastern half of the country. On the West Coast, a semipermanent high-pressure system off the California coast controls the southward movement of storms coming out of the Gulf of Alaska. In the process, the blocking action affects ski conditions in the Sierra Nevada range and keeps most of California rainless all summer.

Another consequence of blocking is its effect on frontal systems. Not only are frontal movements slowed down, but a change also occurs in the makeup of the front, particularly the cold front. Rather than becoming a pronounced line between warm and cold air, the front broadens and often develops a wide band of clouds and rain behind it.

Understanding clouds is essential to successful weathercasting. Of the ten basic cloud types, only three types are likely to produce precipitation of any consequence: nimbostratus, cumulus congestus, and cumulonimbus. Stratus and nimbostratus produce steady precipitation that is light to moderate, while the cumulus clouds produce intermittent and often heavy precipitation. Learning to recognize the various cloud types can thus give a good indication of what today's weather will be like. Another helpful cloud sign is the time when a morning stratus deck "burns off." If it is clear by noon, it means there is a cloud-free sky above the stratus and the remainder of the day will be fair.

A clear sky is also an obviously good indicator of the general weather pattern. But it tells more. It can mean that there is a general descending air motion that indicates a high-pressure system. Or it can mean low humidity. The combination of clear skies and low humidity (together with low winds) indicates that nighttime temperatures will drop sharply.

Clouds can also be reliable messengers that announce a major change in tomorrow's weather. The low-pressure storm systems that typify winter weather patterns are often accompanied by a very predictable sequence of cloud types. As the storm moves in a general easterly direction, the first sign of its impending arrival is trumpeted by cirrus clouds. Three types of cirrus may occur: banded cirrus, mares' tails, and windy cirrus. Windy cirrus almost always are a sign of approaching bad weather. Cirrus are often present about 600 miles or eighteen to thirty-six hours ahead of the storm's warm front.

When cirrus are followed within a few hours by cirrostratus, there is about an 80 percent chance of rain within twenty-four hours. If cirrostratus clouds in turn are followed within several hours by altostratus clouds, there is about a 90 percent probability that precipitation will begin in the next six to twelve hours.

Rain-bearing nimbostratus clouds appear some 200 miles ahead of the warm front. Closer to the front, they thicken and their base becomes lower, while precipitation may become heavier. Rain lasts typically for twelve to twenty-four hours, although it may last longer if the front stalls. The nimbostratus clouds disappear with the passage of the warm front, although not abruptly because the frontal zone is not well defined and may be 50 miles wide.

With the passage of the warm front, precipitation will taper off as winds shift toward the west. Temperatures and humidity can both be expected to rise. Often the air will be unstable enough to form cumulus clouds. These bring the possibility of afternoon showers. This warm region is the buffer zone between the cool air ahead of the warm front and the colder air behind the faster-moving cold front. The warm region can measure hundreds of miles across. Or it may not exist at all at ground level, as in a cold front occlusion.

Unlike the warm front, the cold front usually has a sharp boundary. Skies ahead are frequently clear, with a buildup of nimbostratus and cumulonimbus behind the front. Precipitation may be heavy, although it may only last for several hours. Thereafter, skies will become partly cloudily in the presence of cumulus clouds during the daytime. In the evening, skies generally clear and temperatures fall rapidly.

Sometimes the cold air aloft and behind the front will overtake the warmer air ahead of the front. A very unstable situation then results as cold, dense air rests above the warm, light air near the ground. This can lead to the formation of an intense line of thunderstorms—the *squall line.* Heavy winds and rain (and some-times hail) are a trademark of the squall line. If the temperature differences are large enough, this sets the stage for funnel clouds and tornadoes. Severe weather is covered in detail in Chapter 5.

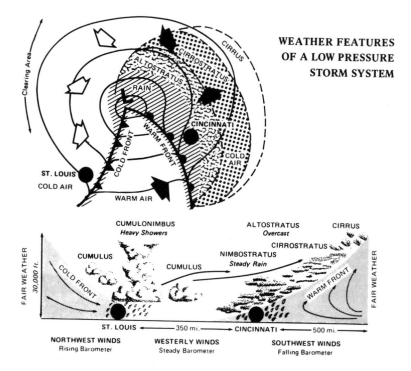

WEATHER FEATURES OF A LOW PRESSURE STORM SYSTEM

Chapter Five

Severe Weather

If outdoorsmen learn nothing else about weather, they must come to recognize, understand, and respect thunderstorms, the lightning that always accompanies them, and the tornadoes they frequently spawn. Thunderstorms present a broad front of dangers that can threaten the lives of unsheltered outdoorsmen. Their gusty winds and the choppy seas they produce are a real hazard to small boats; strong winds and turbulence threaten light aircraft, particularly during landing and takeoff; heavy winds and hail destroy farmers' crops and sometimes the farmers themselves if they are caught unprotected; while the destructive powers of lightning, flash flooding, and tornadic winds are severely potent. Thunderstorms and tornadoes can strike outdoorsmen in virtually any environment, from the mountains to the cities, from the forests to the oceans and lakes. The best way to deal with severe weather is to avoid it. Hikers and campers should take shelter, pilots should not fly in areas of severe weather, and boaters should stay within thirty to sixty minutes of a sheltering shoreline. Keep your eyes on the western sky and know the early warning signs of impending severe weather. And carry and use a weather radio when thunderstorms are possible.

Thunderstorms

Thunderstorms often have their beginnings as small cumulus clouds. But they quickly grow out of this infantile stage, through the adolescent cumulus congestus phase, and into the adult cumulonimbus stage. At maturity, the smaller thunderheads may be only two miles tall, while the larger ones may occasionally reach heights of 10 or 12 miles and span a surface area that can measure from several miles to a dozen or more across. The telltale signs that a full-fledged cumulonimbus has developed are the thunder and lightning and the anvil-shaped, ice-laden top on the cloud; the anvil will always exist atop a mature storm, although it may not be visible because of intervening clouds.

Thunderstorms are nature's way of continually recharging its ionospheric battery. The upper atmosphere has a positive electrical charge in comparison with the ground. Without lightning to transport current from ground to atmosphere, this electrical charge would be neutralized in tens of minutes. At any instant there are an estimated 1,800 to 3,000 thunderstorms occurring simultaneously around the globe. An average day is marked by about 44,000 storms. In the eastern and southern states each thunderstorm lasts an average of twenty-five minutes, although individual storms may only last ten minutes or survive for several hours. The precipitation that falls is similarly short-lived, irregular, and spotty; while one location may get 1 inch of rain in forty-five minutes, a neighboring area a few miles away may experience only blue skies.

Three conditions must be met for a thunderstorm to form:

- The air must be sufficiently humid.
- The atmosphere must be unstable throughout a deep layer.
- There must exist a mechanism to push the lower air layers upward.

Experience has indicated that thunderstorms usually form only when the relative humidity is greater than 75 percent. As air temperatures increase, the same *relative* humidity indicates a greater amount of absolute humidity. As the vertical wind currents push the air upward, the moisture condenses to form raindrops. Moreover, heat is given off in the condensation process. The more moisture that condenses, the greater will be the energy released and the intensity

Lightning strokes during a summer thunderstorm. Courtesy of Noel M. Klein

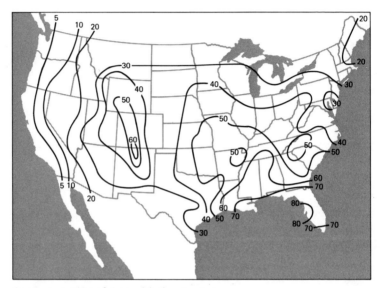

Average number of days with thunderstorms

of the storm. A modest thunderstorm with a diameter of 3 miles contains about one-half million tons of water droplets and ice crystals, and releases some 100 million kilowatt-hours of energy in the condensation process. By comparison, the largest power plant in all of North America (rated at 4,000 megawatts) would need to operate at 100 percent of its capacity for twenty-five hours to produce an equivalent amount of electrical energy.

Given an initial upward thrust, the air will continue to ascend only if the atmosphere is unstable, either by virtue of intense heating at the ground or because of overrunning of colder air aloft. The frequency of thunderstorm activity increases toward the equator for this reason: Not only is the tropical air more humid, but the long days and high sun angles provide plenty of surface heating.

While this leads to thunderstorm peaks in the late afternoon where ground heating is important, most ocean thunderstorms occur during the middle of the night. Surface water temperatures and the temperatures of the overlying air change little from day to night. Higher up, the air does cool off at night—particularly when skies are clear. As the upper-air temperatures drop, the vertical temperature gradient increases and so does its instability.

The third thunderstorm prerequisite—the upward push—is as critical as the need for moist, unstable air. Without an initial shove, the catalyst that starts this thermal engine and keeps it running would be missing. There are three common supplies of this vertical motion: topography, fronts, and air-mass features. Hills and mountains are excellent thunderstorm breeding grounds for this reason. Thunderstorms favor the windward side of mountains because of the steady upward motion. Often thunderstorms will appear to be stationary just to the windward side of the crest or ridge. Actually there is a regular flow of air into and out of the cloud, but the fixed wind pattern causes the cloud cells to continually form and dissipate at the same location. Frontal thunderstorms form in the warm air that is pushed up the surface of either a cold front or a warm front. Because cold fronts have steeper slopes than warm fronts, the vertical air currents, and consequently the storms, are more intense. The altocumulus and nimbostratus clouds that usually mark the trailing edge of the warm front also make it more difficult (if not impossible) to see the thunderheads. Cold frontal thunderstorms are seen more commonly because of the narrower width of the frontal zone and the usual lack of obscuring stratiform clouds. Particularly severe thunderstorms often form in a line or band ahead of a cold front; these squall lines will be discussed a little later in this chapter.

Air-mass thunderstorms get their vertical motion from processes that are neither orographic nor frontal. Typically the upward thrust is caused by the converging action of two wind fields. Florida is a good example. Summer days are marked by on-shore sea breezes on both the Atlantic and Gulf coasts. But these winds do not stop at the shoreline; they penetrate far inland and meet head-on over the center of the peninsula. In the zone where they converge, a regular upward air motion is produced, making the portion of Florida between Tampa and Cape Canaveral the site of the greatest number of thunderstorm-days in the entire country: more than ninety per year.

Near the base of the thunderstorm, air flows into the center of the cloud at gentle speeds of around 2 mph. But at 25,000 feet it is not unusual for the rising air to have accelerated to speeds that approach 60 mph. And the ascending air at the tops of the tallest thunderstorms can reach speeds estimated to be as high as 120 to 150 mph.

Once set into motion, the thunderstorm feeds itself and continues to grow until it reaches a stage where it ultimately begins to destroy itself. Because the rising air moves so fast, it is able to prevent descending raindrops from falling out of the cloud. Instead they are swept along by updrafts and collide with other droplets, growing larger in the process. If the air in the upper reaches of the cloud is cold enough, the droplets will freeze and form hail. This up-down cycle continues as long as the cloud has the energy to retain the ever-enlarging raindrops and hailstones. The start of the rain or hail signals the beginning of the end of the cloud. As the precipitation falls, it drags air along with it, producing the cold down-draft that is characteristic of mature storms. This downdraft severs the top of the thunderstorm from its energy reservoir in the lower levels, thereby effectively strangling the storm. But it is in this final dying stage that the thunderstorm is often most violent, and is marked by lightning, heavy winds, rain, and sometimes hail.

Squall Lines Summer thunderstorms are frequently grouped together in long lines that have come to be called squall lines. More often than not, the damage done by squall lines is much greater than that of a single, isolated thunderstorm. Sheer size is partly to blame: Squall lines are often hundreds of miles long and may

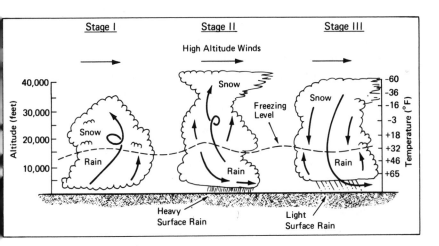

THREE STAGES IN THE LIFE OF A THUNDERSTORM

Stage I. Cumulus clouds form when warm, moist air rises to the point where visible cloud droplets form. In the process, heat is released, causing the cloud to grow even larger.

Stage II. Rain and snow fall inside the cloud. But the strong updrafts in the cumulonimbus catch the raindrops and snowflakes and carry them upward again. Ice particles (hail) form in the cold uppermost reaches of the thunderhead. Heavy precipitation begins to fall out of the cloud; lightning and thunder begin.

Stage III. The downdraft of cold air that accompanies the precipitation signals the impending self-destruction of the cumulonimbus. Updrafts within the cloud cease, and the storm's energy source is cut off. Precipitation weakens, then stops as the downdrafts cease.

be 30 to 40 miles wide. The spacing of the individual thunderstorms that make up the line varies; some are tightly packed in a nearly solid line, while others are separated by 10 to 15 miles. Such groupings also help to intensify the storms as the cold air and the moisture that flow out of one storm help to feed and intensify a neighbor.

Squall lines form on hot, sultry spring and summer days in a variety of locations; the most common place is 100 to 500 miles ahead of and parallel to cold fronts. This type also seems to be the most severe: Prefrontal squall lines frequently consist of several parallel bands (or lines) of thunderstorms whose total width averages 80 to 200 miles, and whose lifespan can range up to twelve to twenty-four hours. Individual storms move northward along the various lines, while the lines move with the prevailing upper-level winds—typically toward the northeast.

Squall lines are most frequent in the middle and eastern sections of the country where a typical summer may bring thirty to an area the size of the upper Mississippi Valley. About two-thirds of these are of the prefrontal variety, while the remainder are equally split between two other types: those found directly along surface weather fronts, and those in the midst of large air masses and not associated with any fronts.

Thunderstorm Weather Violent, severe, intense—thunderstorm weather at its worst is all of these and then some. Unionville, Maryland, has been inundated by a one-minute deluge that measured

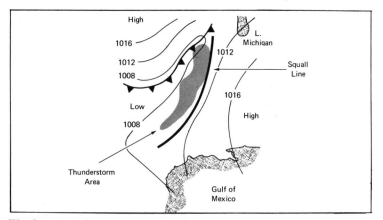

Weather map showing the zone of thunderstorms as detected by radar. *The leading edge of the thunderstorm zone is the squall line and is indicated by the heavy solid line.*

1.23 inches, while Holt, Missouri, holds the world record of 12 inches in forty-two minutes. Hailstones? Potter, Nebraska, was once crushed by grapefruit-sized hailstones weighing up to 1½ pounds.

Thunderstorms do not really sneak up on you. Even if you have no access to radio reports, you normally can have ample forewarning by keeping your eye on the sky *before* the thunderclouds are upon you. Watching the movement of low- and middle-level clouds will indicate the direction and speed of approaching storms. Typically, these speeds are 5 to 20 mph.

The life span of individual thunderstorms is only one or two hours, but as one dies it is often followed by others. So not only must you be on the lookout for approaching storms, you must also be alert to new ones that may form. This is particularly true for air-mass thunderstorms that form away from weather fronts on what begin as apparently fine, sunny days. If puffy cumulus clouds form early in the morning and quickly grow into the towering cumulus congestus stage by ten or eleven o'clock, then take this as a sign that afternoon thunderstorms are a real possibility.

Perhaps the first effect you will feel from an approaching storm is the wind, or lack of it. An isolated and developing storm gently sucks in air at the lower levels over a circular area twelve to fifteen miles across. At the front side of the storm this inflow opposes the prevailing winds, and very light winds may result— the calm before the storm. This usually happens twenty to thirty minutes before the storm matures. Later, in the mature and dissipating stages, a cold and strong downflow occurs in the center of the storm that then rushes out ahead of the storm at speeds of 15 to 25 mph. Its leading edge is called a *gust front,* and it is the location of the strongest and most turbulent surface winds. Wind directions may shift as much as 180° as a storm first approaches and the gust front then passes. This is a particular hazard to boaters and pilots who have to face choppy seas, air turbulence, and strong, ever-changing winds. The gust front is followed one or two minutes later by a rapid drop in temperature of tens of

THUNDERSTORM SAFETY RULES

1. KEEP AN EYE ON THE WEATHER DURING WARM PERIODS AND DURING THE PASSAGE OF COLD FRONTS. When cumulus clouds begin building up and darkening, you are probably in for a thunderstorm. Check the latest weather forecast.
2. KEEP CALM. Thunderstorms are usually of short duration; even squall lines pass in a matter of a few hours. Be cautious, but don't be afraid. Stay indoors and keep informed.
3. KNOW WHAT THE STORM IS DOING. Remember that the mature stage may be marked on the ground by a sudden reversal of wind direction, a noticeable rise in wind speed, and a sharp drop in temperature. Heavy rain, hail, tornadoes, and lightning generally occur only in the mature state of the thunderstorm.
4. CONDITIONS MAY FAVOR TORNADO FORMATION. Tune in your radio or television receiver to determine whether there is a tornado watch or tornado warning out for your area. A tornado watch means tornado formation is likely in the area covered by the watch. A tornado warning means one has been sighted or radar-indicated in your area. If you receive a tornado warning, seek inside shelter in a storm cellar, below ground level, or in reinforced concrete structures; stay away from windows.
5. LIGHTNING IS THE THUNDERSTORM'S WORST KILLER. Stay indoors and away from electrical appliances while the storm is overhead. If lightning catches you outside, remember that it seeks the easiest—not necessarily the shortest—distance between positive and negative centers. Keep yourself lower than the nearest highly conductive object, and maintain a safe distance from it. If the object is a tree, twice its height is considered a safe distance.
6. THUNDERSTORM RAIN MAY PRODUCE FLASH FLOODS. Stay out of dry creek beds during thunderstorms. If you live along a river, listen for flash-flood warnings from the National Weather Service.

SOURCE: National Weather Service

degrees (F). Precipitation begins another few minutes later; it becomes heaviest within two or three minutes and remains so for another fifteen before slowly tapering off.

Where there is smoke, there is fire. And where there is thunder, there is lightning. Lightning is a greater threat to the outdoorsman than tornadoes or hurricanes. Each year lightning kills 150 people in the United States and seriously injures 250. These statistics make lightning 55 percent more deadly than tornadoes and 41 percent more deadly than tornadoes and hurricanes combined.

Lightning

Lightning damage results from three factors. First, the huge electrical current produces a force that can cause nonconducting objects (like trees and bricks) to explode. Second, the current flow also produces intense heat, which frequently starts fires. Third, the shock wave that produces the bang we call thunder also breaks the eardrums of nearby persons. Of the 400 lightning casualties each year, 11 percent are persons who were under trees, 8 percent were on the open water, 7 percent were riding tractors, and 4 percent were playing golf.

Lightning strikes occur worldwide at a rate of 100 every second. They come in many forms: *streaks, forks, broad sheets,* and *ribbons. Heat lightning* is not really a separate form of lightning. It is lightning that can be seen along the horizon, although the storm from which it originates is further away and cannot be seen. *Ball lightning* is perhaps the rarest and most interesting form. It has been reported in the shape of balls and doughnuts ranging in diameter from a few inches to a few feet. It apparently hisses as it moves nearly randomly over the ground, along wires and fences, or through the air. When it disappears, it purportedly often does so by exploding. It has been suggested that sightings of ball lightning have often been reported as close encounters of the first or second kind. It is very understandable that the brilliant, dancing lightball can be reported as a UFO, particularly if the sighting is made on a dark night when objects cannot be seen and used to estimate the true size of the lightning ball.

97

Lightning was first discovered to be an electrical phenomenon around 1750 by Benjamin Franklin in the United States and independently by Mikhail Lomonosov in Russia. Lightning is a giant

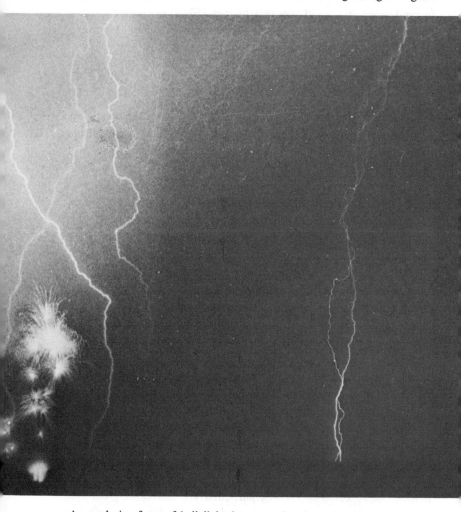

An explosive form of ball lightning occurring in conjunction with streak lightning. Courtesy of Dr. John C. Jensen, Nebraska Wesleyan University

spark that occurs between points or areas that have different electric charges. Most commonly this happens between the base of a thundercloud and the underlying ground, although discharges between different clouds and within a single cloud are not uncommon.

A lightning discharge can only occur once the electric potential exceeds 25,000 volts for every inch that separates the two centers of charge. When this happens, a negatively charged *pilot leader* first advances downward from the cloud base in a rapid sequence of discrete steps. Before reaching the ground it is met by a positively charged *return stroke* that ascends from the ground at lightning speed (one-tenth to one-half the speed of light). The return stroke makes its way up to the cloud by following the highly conducting, ionized trail that was formed fractions of a second earlier by the descending pilot leader. It is the return stroke that we see and identify as the lightning flash. Although it originates at the ground, it appears to originate in the cloud because of the forked path made by the pilot leader that it follows. Once the electrical pathway has been established, additional return strokes follow at fractions of a second until the charge differential has been neutralized— usually less than one-half second. Only about twenty seconds are then needed to reform the cloud-ground charge differential and for another cascade of lightning strokes to occur. The blinding light we see results from light energy given off by highly charged atoms and molecules as they return to their normal, neutral state.

When lightning occurs, tens of thousands of amperes of electrical current flow in a fraction of a second. For comparison, a typical household fuse or circuit breaker will blow when the current exceeds 15 amperes. The heat created by the current flow causes an explosive expansion of the air, which we know as thunder. The distance to an approaching storm can be gauged by noting the difference in time between the lightning flash and thunder clap. Each five-second difference indicates the storm is another mile away (so thunder from a storm 3 miles away will be heard fifteen seconds after the lightning is seen). The length of the lightning stroke may be a mile or more. This means that the rumble of the thunder would last at least five seconds, although echoes and other reverberations allow it to be heard even longer.

Lightning rods are an effective means for actually controlling where lightning will strike. They provide an efficient path for the movement of electrical current between ground and cloud. Because the copper rods that are used are good conductors of electricity, lightning will travel through them rather than other nearby obstacles that are poorer conductors, such as houses and trees. The height of the lightning rod is very important; the higher, the better. A 15-foot-tall copper rod will not be very effective if 25-foot houses surround it. Lightning prefers to seek out the tallest and best-conducting obstacle. Unfortunately, that can be the putter in your hand as you stand with arms raised after parring the third hole on an exposed, flat golf course. In cities, tall buildings are particularly prone to lightning strikes. New York's Empire State Building receives an average of 48 per year, while Mount San Salvatore near Lugano, Switzerland, averages 100.

To avoid becoming a lightning statistic, do not stand on, under, or near tall hills, towers, flagpoles, or trees. As far as lightning is concerned, height is relative. A small tree on the plains of Nebraska is as likely to attract a strike as the Gateway Arch in downtown St. Louis. In the open, a safe rule is to be at least as far away from a potential lightning rod as the height of that obstacle. This means, for example, putting at least 100 feet between yourself and a nearby isolated 100-foot-tall tree. If possible, stay even further away and seek out a depression in the terrain and lie down until the storm has vented its fury or moved safely away.

Sometimes it is impossible to move away from possible lightning strikes. If you are in a car, stay inside, as the vehicle is a good insulating compartment. Sailing or boating on open waters presents a distinct hazard. If your boat does not have lightning protection, it is crucial that you jury-rig a system in the face of an approaching electrical storm.

Experts suggest that even quite primitive methods can be fairly effective. The guiding principle is that you establish a metallic path between the highest point on the boat and the water. Heavy copper wire (#8 or heavier) is best, but nearly anything will provide at least some protection. The top of your lightning rod must be high enough to blanket all of the boat within a 45° cone (that is, the horizontal distance between the lightning conductor and any part of the boat must be less than the height of the conductor above the deck). The mast on a sailboat makes an ideal support for a lightning rod. If the top of the lightning rod is not tall enough, then it can be extended by lashing a pole to it, or it may be necessary to use multiple rods to blanket the entire vessel. The conducting rod or wire must be "grounded" in the water by attaching a large metallic plate (at least one foot square) to improve electrical contact with the water. Do not touch any part of the conducting wire during an electrical storm, and be sure that it is isolated from any part of the boat that is also an electrical conductor. As a last precautionary measure, be absolutely certain that your body never makes contact simultaneously with two different pieces of your boat's gear that are potential electrical conductors (for instance, winches, radio, fathometer); you could become an electrical conductor yourself.

Pausing for a moment from the dangerous aspects of lightning, it is perhaps equally beneficial that we enjoy the beauty of nature's very own light show. Not only is it impressive at the time, but it can be enjoyed again and again in the form of lightning photographs. High-resolution photos that show intimate details of individual lightning strokes require sophisticated equipment, patience, experience, and good fortune. But acceptable-quality photos are readily achievable with an ordinary camera plus some patience and a little luck. Two techniques are available: short exposure and time exposure. For a photo that captures many lightning strokes, set the camera on a tripod, point it in the general direction of the storm, open the shutter and close it only after lightning

LIGHTNING SAFETY RULES

THESE SAFETY RULES WILL HELP YOU SAVE YOUR LIFE WHEN LIGHTNING THREATENS.

1. Stay indoors, and don't venture outside unless absolutely necessary.
2. Stay away from open doors and windows, fireplaces, radiators, stoves, metal pipes, sinks, and plug-in electrical appliances.
3. Don't use plug-in electrical equipment like hair dryers, electric toothbrushes, or electric razors during the storm.
4. Don't use the telephone during the storm—lightning may strike telephone lines outside.
5. Don't take laundry off the clothesline.
6. Don't work on fences, telephone or power lines, pipelines, or structural steel fabrication.
7. Don't use metal objects like fishing rods and golf clubs. Golfers wearing cleated shoes are particularly good lightning rods.
8. Don't handle flammable materials in open containers.
9. Stop tractor work, especially when the tractor is pulling metal equipment, and dismount. Tractors and other implements in metallic contact with the ground are often struck by lightning.
10. Get out of the water and off small boats.
11. Stay in your automobile if you are traveling. Automobiles offer excellent lightning protection.
12. Seek shelter in buildings. If no buildings are available, your best protection is a cave, ditch, canyon, or under high-high clumps of trees in open forest glades.
13. When there is no shelter, avoid the highest object in the area. If only isolated trees are nearby, your best protection is to crouch in the open, keeping twice as far away from isolated trees as the trees are high.
14. Avoid hilltops, open spaces, wire fences, metal clothes lines, exposed sheds, and any electrically conductive elevated objects.
15. When you feel the electrical charge—if your hair stands on end or your skin tingles—lightning may be about to strike you. Drop to the ground immediately.

Persons struck by lightning receive a severe electrical shock and may be burned, but they carry no electrical charge and can be handled safely. A person "killed" by lightning can often be revived by prompt mouth-to-mouth resuscitation, cardiac massage, and prolonged artificial respiration. In a group struck by lightning, the apparently dead should be treated first; those who show vital signs will probably recover spontaneously, although burns and other injuries may require treatment. Recovery from lightning strikes is usually complete except for possible impairment or loss of sight or hearing.*

*See H. B. Taussig, "Death from Lightning and the Possibility of Living Again," *Annals of Internal Medicine,* vol. 68, no. 6, June 1968.
SOURCE: National Weather Service

occurs in the field of view. Your wait will only be about one-half minute during an active storm. Using ASA 100 film, the aperture should be set to f4.5; focus the lens at infinity or a trifle closer. More resolution is possible with less chance of the film fogging if the short-exposure method is used; your chances for success are not as good and the amount of film consumed is greater, but the quality is apt to be better. First, set the shutter speed at $\frac{1}{15}$ or $\frac{1}{30}$ second. Then, at the instant a stroke is seen, snap the shutter. You will not capture the stroke that triggered your reflex, but you have a fair chance of photographing one of the subsequent strokes that usually follow within fractions of a second. Unfortunately, it is the first stroke that usually has the intricate branches; the subsequent strokes normally are unbranched.

Tornadoes

The severe thunderstorm can be deadly in yet another way, for it is the father of the tornado. Tornadoes are funnel-shaped clouds that are usually less than a mile in diameter and that move forward at a speed of around 40 mph. Their winds spiral in a counter-clockwise direction at speeds as high as 300 to 400 mph, which is fast enough to pick up and move an 83-ton railroad car 80 feet through the air and drive a piece of straw through a 2-by-4-inch block of wood. Their path of destruction is typically only a quarter-mile wide and 16 miles long. Yet a 1917 tornado ravaged 293 miles across Illinois and Indiana over a seven-hour period.

The presence of a severe thunderstorm is the first stage in the development of a tornado. These thunderheads are most often found in a squall line, and can sometimes be identified by a formation of breastlike *(mammata)* clouds hanging from the base of the thunderstorm. Next, a funnel cloud develops at the base of the thunderstorm cloud. The funnel is visible because of the water droplets and dust it contains. Sometimes the funnel cloud twists like a snake as it descends from the thundercloud to the ground. Only

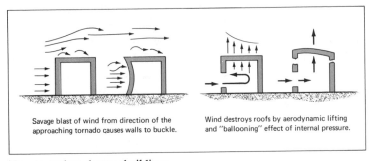

Savage blast of wind from direction of the approaching tornado causes walls to buckle.

Wind destroys roofs by aerodynamic lifting and "ballooning" effect of internal pressure.

How tornadoes destroy buildings

when it has touched the ground is it called a tornado. (Over water it is called a *water spout.*) When the funnel reaches the ground, it usually darkens from the debris it sucks up. One characteristic of all tornadoes is the tremendously loud noise that accompanies them. No one knows exactly what causes the noise, but witnesses have described it as the "roar of a thousand trains" or the "buzzing of a million bees" or the "thunder of a fleet of jet airplanes."

The tornado displays its awesome destructive power in two ways. The exceptionally strong and gusty winds can easily topple cars and trees. Coupled with the winds, the partial vacuum at the center of the tornado lifts objects off the ground and tosses them about like toys. It can cause houses to literally explode because of the pressure difference between the sealed building and the tornado core.

Not all tornadoes are alike, although all are potential killers. The National Weather Service (NWS) has identified three broad categories of tornadoes. The weakest of all is the "mini-tornado" which has winds of less than 100 mph. It may be up to 100 feet in diameter and last only a few minutes, traveling up to one-half mile. Because it is so small, a mini-tornado may not be predicted by an official "tornado watch," nor may it be reported in a "tornado warning." A tornado watch is an alert issued by the NWS that a tornado or severe thunderstorm is possible in a specified area;

watches are normally given for rectangular areas 140 by 200 miles. It is essentially an advisory to be aware of the possibility that severe weather may develop in the watch area. The tornado warning is an alert that is given only when a tornado has actually been sighted. The warning specifies where and when the tornado is expected to strike downwind areas.

The "medium tornado" stands a better chance of being accurately predicted and observed. It is up to 400 feet across, with winds up to 150 mph. Typically, it lasts up to twenty minutes and can travel up to 10 miles in that time. The most destructive category is the "maxi-tornado"; it also has a very good chance of being predicted and observed. Most tornado-caused deaths are the result of maxi-tornadoes. Up to 1½ miles wide, they typically travel up

Mobile-home park in Willard, Ohio, destroyed by a tornado. Courtesy of the Ohio National Guard

to 200 miles in a three-hour period. Peak wind speeds range from 150 to 300 mph.

About 650 tornadoes occur each year in the United States. Twenty-five percent of all tornadoes occur between 4 P.M. and 6 P.M., while 83 percent occur between noon and midnight. About half occur during April, May, and June. During March and April, tornadoes are generally more common along the Gulf Coast. They then shift northward over the next few months to threaten Kansas, Iowa, and Nebraska in June. There is then a tendency for them to occur farther eastward. The loss of human lives averages about 120 per year in the United States. A "quiet" year was 1961, with 51 tornado deaths, while during the infamous Palm Sunday tornadoes of 1965, 299 deaths were recorded. Property damage from tornadoes also varies greatly from year to year: In 1961 property damage was around $68 million, while the total for 1965 was $400 million.

TORNADOES AND TORNADO DEATHS BY STATE (1953-1970)

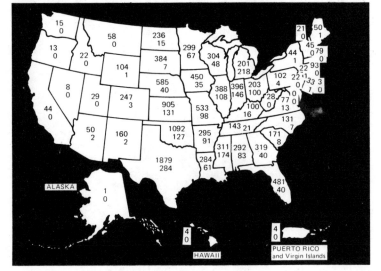

Upper figure is number of tornadoes   *Lower figure is number of deaths*

TORNADO SAFETY RULES

When a tornado approaches, your immediate action may mean life or death. Seek inside shelter, preferably in a tornado cellar, an underground excavation, or a steel-framed or reinforced concrete building of substantial construction. Stay away from windows!

IN CITIES OR TOWNS

Office Buildings

Go to an interior hallway on the lowest floor, or to the designated shelter area.

Factories

Workers should move quickly to the section of the plant offering the greatest protection in accordance with advance plans.

Shopping Centers

Go to the designated shelter area—not to your parked car.

Homes

The basement offers the greatest safety. Seek shelter under sturdy furniture if possible. In homes without basements, take cover in the center part of the house, on the lowest floor, in a small room such as a closet or bathroom, or under sturdy furniture. Keep some windows open, but stay away from them.

Mobile Homes

Mobile homes are particularly vulnerable to destructive winds. Proper tie-downs to prevent overturning will minimize damage. A warden should be appointed in mobile-home parks to scan the skies and listen to radio and television warnings. There should be a designated community shelter where residents can assemble during a tornado warning. If there is no such shelter, do not stay in a mobile home during a tornado warning. Seek refuge in a sturdy building or a ditch, culvert, or ravine.

Schools

Whenever possible, follow advance plans to an interior hallway or the lowest floor. Avoid auditoriums and gymnasiums or other structures with wide, free-span roofs. If a building is not of reinforced construction, go quickly to a nearby reinforced building or to a ravine or open ditch and lie flat.

TORNADO SAFETY RULES *(continued)*

OPEN COUNTRY

If there is no time to find a suitable shelter, lie flat in the nearest depression, such as a ditch or ravine.

KEEP LISTENING

Your radio and television stations will broadcast the latest National Weather Service tornado watches and warnings, and inform you when the danger is over.

WATCH THE SKY

Tornadoes come and go so quickly that there may not be time for a warning. During a tornado watch, be alert for the sudden appearance of violent wind, rain, hail, or funnel-shaped cloud. When in doubt, take cover. Tornadoes are often obscured by rain or dust. Some occur at night.

REMEMBER

Tornado Watch means tornadoes are expected to develop. *Tornado Warning* means a tornado has actually been spotted. Persons close to the storm should take cover immediately. Those farther away should take cover if threatening conditions approach.

PART TWO

Weather Where You Are

Chapter Six

Lakes and Oceans

Sailors, fishermen, and powerboaters face similar hazards on the water: Fog, wind, waves, and lightning are perhaps the four that are most important. Here we will concentrate on fog, wind, and waves (lightning hazards and precautions were discussed in Chapter 5). Fog is a critical weather element on the water because it impairs safe navigation and because it occurs more frequently over water than land. Wind affects boating two ways: one, direct and the other, indirect. Winds affect boaters indirectly through the waves they create. The force of the wind acts directly on boats and boaters, hurling spray or propelling sailboats on a reach, close-hauled, or on a run. Powerboats see the wind as foe rather than friend, drifting as a result of the force exerted by the wind on the parts of the boat that rise above the water line.

While much of the material covered in this chapter is most relevant to boaters, fishermen, and mariners of all types, it is equally germane to landlubbers who find themselves on the shores of the coastal mainland or on relatively small islands (no more than a mile or two across). The weather on these land areas will differ little from conditions on the water.

Fog

Fog has many causes and more numerous effects. Walking a fog-shrouded beach can be peaceful and relaxing amid the beauty and serenity of the muted colors and muffled sounds. Usually the wind is nothing more than a gentle breeze and the sea is quiet in the absence of breakers. But 100 yards offshore the mood can change abruptly. Visibility is sharply reduced, land-based navigational aids are hidden, and the bearing of foghorns often cannot be resolved accurately. Radar is of course an invaluable aid in fog, yet collisions even among radar-equipped boats and ships are not uncommon.

Understanding the nature and causes of fog is essential. If the mariner knows where and when fog is likely to occur, he can often avoid it, either by plotting a different course or sailing at another time. My own first exposure to dense fog is a good example. It was June, and our 450-foot ship had anchored off Martha's Vineyard for two or three days of boat drills. But rather than pull an oar in a monomoy, I spent a good many hours on the fantail sounding a fog bell in soup so thick I couldn't make out the ship's bridge less than 100 yards away. After a few days the problem was solved: We changed the weather! Moving 50 miles south, where the warmer surface waters off Long Island's north shore inhibited fog formation, we increased our visibility to several miles in the same stagnant air mass.

What Is Fog? Fog is a cloud consisting of water droplets so small that they remain suspended in the air and reduce visibility at the earth's surface to less than 1 km (.62 mile). A cloud is not truly fog if its base does not reach to the sea or ground surface, or if visibility is greater than 1 km.

There are five different types of fog you are likely to encounter on the water. Each is formed in its own unique way, although in all cases both of the following must occur: (1) The temperature and dew point of the air must be equal, and (2) the air must contain tiny particles on which the water vapor can condense into droplets. These particles are called *condensation nuclei* and are particularly abundant near salt water (in the form of evaporated salt particles) and cities (because of particles from industrial wastes).

There are two different ways in which the air can become saturated. The first, and more common, mechanism is for the air to be cooled sufficiently so that the temperature drops to the dew point; *radiation fog, advection fog,* and *up-slope fog* are formed in this way. The second, less common mechanism is for the dew point to be raised to the air temperature by the addition of water vapor; *steam fog* and *frontal fog* are produced in this way.

Advection Fog By far the most important variety found at sea, advection fogs comprise four marine fogs in five. Advection refers to horizontal movement or transport that takes relatively warm and humid air to a cooler location to form fog. Perhaps the most famous example occurs near the Grand Banks east of Newfoundland. Here, air that has first moved across the warm Gulf Stream (picking up moisture by evaporation from the sea surface along the way) encounters the cold waters of the Labrador Current. The warm, humid air in contact with the cold surface quickly cools to the dew point to produce fog. When the wind blows at a moderate speed (in the range of 4 to 12 mph), the turbulence in the air mixes the cool, fog-laden air upward to create a thick fog layer. Slower winds confine the fog to a shallow layer at the sea surface, while higher winds rapidly dissipate the fog and improve visibility. The difference in air and sea surface temperatures is often so great at the boundary between the warm and cold ocean currents that a virtual fog wall is produced. Other places famous for their advection fogs include the British Isles, where air from the Gulf Stream encounters the cool land area; the Aleutian Islands, meeting grounds for the warm Japan Current and the southerly outflow of cold air and water through the Bering Strait; and the California-Oregon coast, where cold bottom waters are brought to the surface by the California Current to chill the prevailing onshore flow of Pacific air. Advection fogs are also frequently formed in autumn

Advection fog at sea. Courtesy of Daniel L. Haynes

and winter along the coasts of large lakes and oceans when the warm, humid marine air moves over the colder ground surface. In this situation both coastal and inland navigation can be impaired: Inland, rivers and lakes are engulfed in fog; offshore, coastal navigation aids may be obscured.

Fortunately, advection fog is fairly easily forecast over water. First, daytime air temperatures must be warmer than the sea surface temperature. Second, the dew point computed during the day must also be higher than the sea surface temperature. If both of these conditions are met and if the winds are less than 10 mph, then advection fog will form as the lowest layers of air are cooled to the temperature of the water surface. This is particularly probable in the evening, when the air is not warmed by the sun. The chart on page 116 provides a step-by-step procedure for predicting when cooling will produce fog. Applying the procedure in reverse allows you to estimate what conditions are necessary to dissipate a fog.

Radiation Fog and Upslope Fog Both occur only over land, yet they may also impair safe navigation by obscuring coastal aids to navigation. Radiation fog occurs at night when clear skies allow the ground surface to lose heat rapidly by giving off infrared radiation. As the ground cools, so does the air in contact with it. When the air is calm, the lowest fractions of an inch of air may cool to the dew point, and dew will form on the ground. But if the cooling is great enough and the air is mixed by a gentle breeze, a layer of radiation fog will form. Upslope fog, on the other hand, results from the cooling that occurs as air rises to flow over the terrain (such as coastal bluffs and hills). If the air is already sufficiently humid, then the cooling of 1°F per 180 feet of rise may be sufficient to produce a thin fog or to thicken an existing advection fog.

Frontal and Steam Fog The two types of fog that form by the addition of water vapor are both less common and less important to the boater than cooling fogs. Frontal fog can form along weather

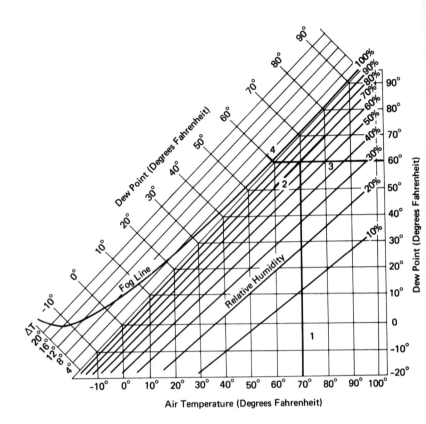

Air Temperature (Degrees Fahrenheit)

FOG FORECASTING CHART

• *The first thing you need to know to make your own fog forecast is the dew-point temperature. Find the point where the (vertical) air-temperature line crosses the (slanted) relative-humidity line. The dew point is then read directly as the value of the horizontal line that passes through this point.*

• *Fog will start to form when the air temperature has cooled to the dew point. But further cooling of the air is necessary to create enough water droplets to reduce the visibility to 1 km—the definition for fog.*

• *The additional cooling needed is the distance along the diagonal "$\triangle T$" lines from the 100% relative-humidity line to the curved "fog line."*

(Continued on next page)

fronts when relatively warm raindrops fall through layers of cooler, humid air. As they fall, the droplets practically evaporate and in the process may raise the dew point to the temperature of the air. The vapor then condenses to form a layer of minute water droplets that occasionally extend downward to the earth's surface.

Steam fog, the second type produced by the addition of water vapor, is worth noting not for its impact on navigation but rather for its chilled beauty. Steam fog forms when very cold air flows over comparatively warm water. This can occur during the autumn and early winter when "zero-degree" arctic air passes over warm ocean currents. Three things happen: (1) Evaporation occurs at an intense rate from the warm water surface; (2) the air in contact with the water warms and becomes unstable; and (3) the moist, unstable air rises to mix with the cold air above and condense. The end product, also called *sea smoke* or *arctic smoke,* has a wispy appearance not unlike that of cotton candy. Interesting to watch is the way the "smoke" writhes and dances above the water, reflecting the twisting and turning of the turbulent eddies in the air.

• *The example below is numbered on the diagram.*
(1) The noontime temperature is 70°F.
(2) The noontime value of relative humidity is 70%.
(3) Therefore, the dew point is 60°F.
(4) The additional cooling required to form fog is 1°F. Therefore, fog will form if the temperature of the air mass drops to 59°F. A good rule of thumb is that over the water the nighttime air temperature will approach the temperature of the water surface. So, if the water is warmer than the fog-formation temperature (59° in the example), you can forget about fog. But if the water temperature is lower than the fog-formation temperature, then fog can be expected.

TABLE 14. FOG RECORDS FOR THE UNITED STATES	
Foggiest place over a 10-year period	Cape Disappointment, Washington: 2,552 hours per year (29% of all hours)
Foggiest place over a 4-year period	Willapa, Washington: 3,863 hours per year (44% of all hours)
Foggiest place over a 1-year period	Willapa, Washington: 7,613 hours (87% of all hours, or an average of 20 hours and 52 minutes per day!)

Local Winds

The wind at any instant can be the product of many forces and effects, particularly on lakes and coastal waterways where local conditions can either modify the prevailing winds that accompany large-scale weather systems or create wind where none would otherwise occur. Local wind effects have two causes: the *aerodynamic* influence of the terrain, and the difference in the *thermal* properties of land and water. The aerodynamic effect of local topographic features can be extremely important. In fact, the world's record ground-level wind speed (excluding tornadoes) was produced by local terrain channeling at the Mt. Washington weather observatory where a wind gust of 231 mph and a five-minute average speed of 188 mph were recorded on April 12, 1934. Similar effects occur on a broader if less dramatic scale along many coastal reaches, such as the Golden Gate and Carquinez Strait outside San Francisco. Experienced mariners recognize the intricacies of these local wind circulations in their home waters; understanding their causes can help avoid unpleasant surprises when sailing less familiar waters.

Aerodynamic Effects The concept of local aerodynamic wind effects from hills, straits, islands, and so forth is quite simple: Air driven by large-scale forces encounters a topographic obstruction and is then diverted either around or over the obstacle. Because the terrain restricts the space through which the air can flow, the speed of the wind is increased accordingly so that the total volume of air passing through the area remains the same. However, sometimes most of the air flows *over* the obstruction and relatively little goes *around,* while at other times the reverse is true. Why? The answer lies in the stability of the air: Unstable air accelerates in the direction it is pushed; neutral air moves as pushed as long as the pushing force is present; stable air resists any vertical shove and seeks to return to its original height. But how does the boater estimate stability without the benefit of sophisticated instruments to measure temperatures aloft? Although not infallible, several rules are quite reliable. When the prevailing wind flow is offshore, sunny days are usually unstable (especially when cumulus clouds abound), clear nights are stable, and sunrise and sunset periods are neutral. Air that has blown 10 to 20 miles or more over water is unstable

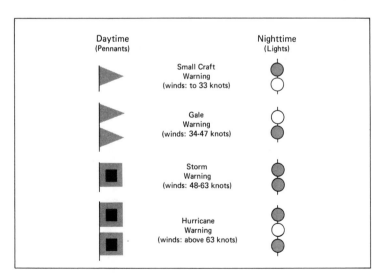

Marine warning signals

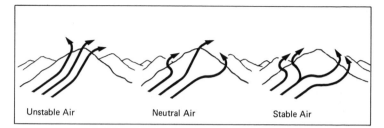

Unstable Air   Neutral Air   Stable Air

Airflow over and around hills and mountains depends on the stability of the air.

when water temperatures are higher than air temperatures, and stable when the water surface is cooler. Skies overcast with sheetlike stratus clouds are very nearly neutral all of the time.

The aerodynamic effects the mariner must contend with are identical with those the outdoorsman faces on land, except for a few simplifications that occasionally exist. With air that has had a long fetch over water, the stability typically will be neutral or nearly so—the first simplification. If some upwind islands are small (say, less than 1 mile in diameter), or if they are larger but the sky is overcast or the wind is strong, then you can ignore the thermal effects the islands will have on the wind flow—the second simplification. What we have left is not an uncommon situation for the mariner: being in close proximity to an isolated, nearly circular island and having to contend with effects on the wind flow and subsequently on the operation of the craft.

As the wind approaches, it begins its diversion around and over the island about one to two island-widths upwind. To either side of the island, the wind speed increases dramatically to a maximum that is nearly twice the undisturbed wind speed. Upwind, the winds drop off considerably. But the most dramatic changes occur downwind. There the air that has traveled over the island is sucked back toward the island at water level. This downward flow of air hits the water downwind at a distance that is one or so times the height of the island; it usually can be seen by the turbulent wave or chop pattern it makes on the water. At the same time, the (horizontal) wind that has flowed around the island also recurves toward the leeward side of the island. Together, these two effects create a very turbulent and gusty zone that extends downwind of the island from about four o'clock to eight o'clock, where the clock face represents a bird's-eye view of the island and the wind approaches from twelve o'clock. In this sector the winds change their direction rapidly, while speeds range from near-calm to twice the undisturbed wind speed. The turbulent zone diminishes in intensity further downwind, but a noticeable effect is still present five to ten diameters downwind.

This only scratches the surface of the aerodynamic effects confronting the mariner in coastal waters, near islands, or on rivers and small lakes. Other aerodynamic wind effects are the subject of Chapter 7. Thermal effects usually are found in conjunction with aerodynamic effects, although not always. The following section focuses on the thermal effects that are most important and familiar to the mariner—the diurnal cycle of land and sea (or lake) breezes.

Thermal Effects Differences in the thermal properties of land and water are the cause of local, thermal-type winds: the familiar *land* and *lake/sea breezes.* When the sun rises, early morning ground and air temperatures rise very rapidly—as much as 4° or 5°F every hour. At the same time, water temperatures barely change. As a result, two dissimilar mini-air masses form: one over land, the other over water. Differing pressures, humidities, and temperatures cause a low-level movement of air from water to land during the daytime (the sea or lake breeze), and from land to water at night (the land breeze). Higher up in the atmosphere, the winds flow in opposite directions to complete the cycle.

To understand the weather created by land and sea breezes, it will be helpful to understand the hows and whys of these thermal winds. Let's start with a simple setting: Our ocean shoreline is

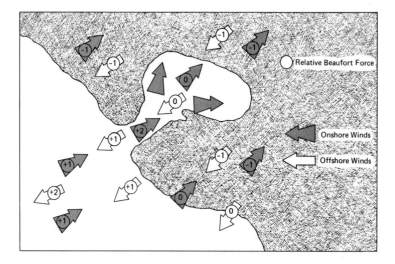

The change in wind speed over land, in a harbor, and at sea. *Judging the speed of the wind from an inland location will require you to make a correction to accurately estimate the wind in a nearby harbor and at sea. Wind blowing over land is slowed down by friction. As the air encounters the smooth, open harbor waters, it speeds up—about one Beaufort force's worth. And when it next blows out over an extended body of water, it has the fetch necessary to "forget" its rough surface history and speed up to its full potential—one to two Beaufort categories above its speed in the harbor. The picture is somewhat different when the wind blows onshore. As it meets the rough land surface, it almost immediately decreases in force one category at the shoreline, and then farther inland. In the narrow harbor entrance, channeling may first increase the speed one category before the wind decreases in the harbor to one force category below its undisturbed value. For example, if the wind at sea is Force 5, it will be Force 6 in the channel and Force 4 in the harbor. See pages 18 and 19 for the tabulation of Beaufort wind forces categories.*

at the center of a large air mass, and the large-scale winds are calm. Sea-level pressure and its decrease with altitude are initially identical over water and land. It's early summer, and by early morning air temperatures over land are already 5°F warmer than at sea. The density of the ground-level air over land decreases as the temperature increases, but the ground-level pressure stays constant because the weight of the *total* amount of air overhead is unchanged. What has happened is that the lower air layers above the land surface have expanded as they warmed and have pushed the upper air layers further upward. Nothing has changed over the water. But if we measure the air pressures that now exist at about 1,000 to 3,000 feet, we see that the pressure there is *greater* over land than water. In an attempt to equalize pressures, upper-level air moves from the land to the water; and by adding more air to the ocean mini–air mass, sea-level pressures rise over the water surface as ground-level pressures fall over land.* Now it is the turn of the lower air layers to move away from the higher-pressure ocean air. This is the sea breeze that begins to blow by midmorning along the coast (usually around nine o'clock).

The sea breeze strengthens as the land-water temperature difference increases, and will normally continue to blow for some hours past sunset (although with diminishing speed). During the night and the early morning hours, the entire process reverses itself. Water temperatures become warmer by comparison with the rapidly cooling land surface. The resulting land breeze blows from land to water in the lower layers, while 500 to 1,500 feet above it blows onshore.

Land and sea breezes form when land and water temperatures differ significantly—usually by 10° to 15°F and more. This fact

---

*The pressure difference between land and water is only .5 to 1 millibars (.015 to .03 in. Hg), but the small distance makes the pressure gradient quite large (see Chapter 1, page 12).

alone suggests some guidelines for estimating when breezes are likely to occur and how strong they will be:

- As the large-scale winds strengthen, the identity and effect of the sea breeze are camouflaged.
- Advection of cold air by large-scale air masses will overwhelm and destroy local heating patterns (spring and summer are the common sea breeze seasons).
- Cold offshore currents (particularly with the upwelling of cold bottom waters) favor sea breeze development.
- As the cloud cover increases, the strength and frequency of the sea breeze decrease.
- High humidities retard the nocturnal cooling of the land surface; the sea breeze will persist longer into the night, while the land breeze will be either weakened or eliminated.
- Wet ground surface inhibits both sea breeze and land breeze formation by keeping daytime temperatures down and nighttime temperatures up.

The sea breeze (at the surface, that is) is felt initially at the shoreline and then further inland and offshore. The leading edge of the sea breeze is called the *sea breeze front* and marks the boundary between the cool, moist maritime air and the warm, dry land air. The sea breeze front typically moves inland at speeds of 3 to 7 mph. The distance inland to which the sea breeze can penetrate is governed by the size of the water body and the topographic features of the coastline, in addition to the difference between land and water temperatures and the number of daylight hours. In the tropics the sea breeze can be felt up to 90 miles inland from the ocean, while in temperate latitudes the sea breeze will only be felt 10 to 30 miles from the coast. The intensity of the wind (that is, the speed of the sea breeze) also varies. Small lakes will create lake breezes of up to about 5 mph, while large lakes and seacoasts will typically induce winds of up to 8 to 12 mph in the United States. Where inland heating is intense and

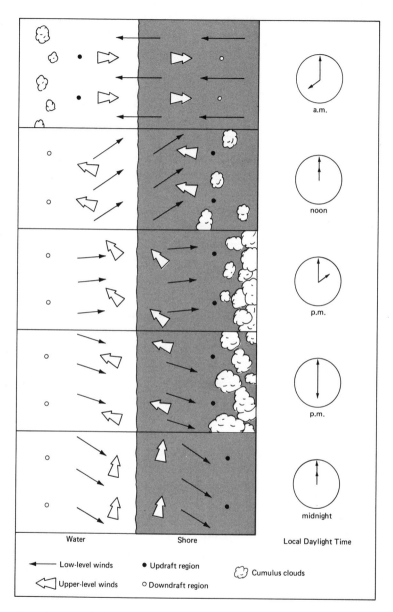

Water   Shore   Local Daylight Time

— Low-level winds   ● Updraft region   ☁ Cumulus clouds

◁ Upper-level winds   ○ Downdraft region

Five phases of the land/sea breeze cycle

ocean waters are cool, sea breezes of up to 20 to 25 mph are not uncommon. The Monterey-to-Santa Cruz coastline is one such location; inland temperatures are commonly in the range of 90° to 100°F during summer, while the southward-flowing California Current brings cold bottom waters to the surface, creating 40°F+ ocean/land temperature differences.

The Coriolis force (see Chapter 1) has an important influence on the sea breeze. The longer it takes for air to move from water to land, the more pronounced is the Coriolis effect that causes the sea breeze to turn clockwise (or to *veer)* along its fetch or path. Theoretically, if the sea breeze blew long and far enough, and if the land were as smooth as the water, the sea breeze would eventually blow parallel to the shoreline. This does not actually happen, although the wind direction will usually shift throughout the midday hours until it makes about a 30° to 45° angle with the shoreline. The roughness of the land surface acts on the wind in much the same way that a soft shoulder pulls a car off the highway. By forcing the wind toward the rough land, the sea breeze is inhibited from turning further clockwise.

Apart from the wind it creates, the sea breeze produces additional effects inland. As the sea breeze front marches inland, it brings with it air that is both cooler and more humid. Drops in temperature can vary from 1° or 2°F up to 5°F in a matter of minutes, while the humidity increases simultaneously by 10 to 30 percent. With surface sea breeze blowing onshore (up to heights of 1,500 to 3,000 feet) and the overriding air layer (about half the thickness of the lower layer) blowing offshore, there are corresponding regions of upward and downward air motion. The daytime motion over the water is downward and results in generally clear skies and no precipitation. Over land, the moist ocean air rises and creates fluffy cumulus and sometimes cumulus congestus clouds. The clouds form in a line parallel to the shore and frequently produce nominal amounts of precipitation. In the period from late night to early morning, the wind flows are reversed and it is common to see a line of cumulus clouds over the water. Because the nighttime land breeze is less intense, precipitation is not typical from the cumulus clouds that form.

The nighttime land breeze is less pronounced in other ways. The land breeze normally does not form until sometime around midnight, and is gone shortly after sunrise. Wind speeds are about the same as the more common sea breeze speeds, except that the more extreme speeds do not occur.

Sea and land breezes are an important feature of coastal weather. For centuries, fishermen have been able to use the land breeze to go to sea at night and have returned to shore with the morning sea breeze. Recognizing their cycles, outdoorsmen can use them to advantage by anticipating the weather changes they bring—both good and bad. If you know when the large-scale weather pattern will favor the formation of sea breezes, you stand an excellent chance of forecasting clouds, winds, temperatures, and humidity throughout the day. When large-scale winds are present, you should consider the effect they will have on local conditions and the sea breeze cycle. Strong winds (generally greater than 15 mph) will overpower the sea breeze and prevent it from forming. Weaker large-scale winds, on the other hand, will modify the sea breeze so that the actual winds will be a combination of the large-scale and local winds.

Wind and Waves

Wind affects a boat's movement in two ways: It develops wind waves, and it exerts a drag directly on the boat. The amount of drifting is directly proportional to the speed of the wind and the cross-sectional area that the boat presents to the wind. By heading directly into the wind, this frontal area and the power needed to maintain steerageway are minimized. However, it is also necessary to consider the effects of waves on the navigation of the boat and the fact that the wind and wave directions usually differ by one or more points of the compass (one compass point = 22.5°); thus the usual procedure is to maintain a heading about one compass point to the movement of the wave front.

The most important effect of the wind on boating is its role in generating wind waves. *Wind waves* represent only one of three types of water waves; the other two are *gravitational waves* and *earthquake waves.* The lunar tides that occur about every twelve hours are called gravity waves or true tidal waves. Waves caused by the rapid movement of large masses of earth are earthquake or seismic waves or tsunamis, although they are commonly misnamed tidal waves. Earthquake waves have periods that range from fifteen to sixty minutes and can do great damage by inundating low-lying coastal areas with wave heights of up to 50 feet in shallow coastal waters (in deep water, the wave height may be but 1 or 2 feet). But even the earthquake wave does not dwarf the highest wind-driven wave, reported to have had a height of 112 feet.

Waves have three characteristic features: (1) *wave height*—the vertical distance from the top or *crest* of the wave to its low point or *trough;* (2) *wavelength*—the horizontal distance between successive wave crests; and (3) *wave period*—the time it takes a wave crest to travel one wavelength. The four common and distinct wind-wave types are conveniently classified by their periods:

Type of Wind Wave	Typical Period
Ripples	Less than 1 second
Wind chop	1–4 seconds
Fully developed seas	5–12 seconds
Swell	6–16+ seconds

The simple mental picture one gets of each of these wave types often bears little or no resemblance to the actual state of the sea. This is not to say that these separate wave types do not exist. They do, but usually in combination rather than alone. And several wave directions often coexist. The swell from a storm that passed a few days earlier may intermingle with wind chop from an afternoon sea breeze. A swell is composed of "old waves" that have left the wind source that created them. In addition to their longer period, swells have very long wavelengths and relatively flat, low crests.

How do wind waves develop and grow in the first place? They start with the atmospheric wind which is gusty and has small-scale pressure fluctuations. As some of the energy from the wind flow is given to the water, ripples appear on the surface. Ripples present a sloping surface to the wind, allowing the wind to push directly on the water. In the process, energy transfer from air to water is quicker and more efficient, and the waves grow higher. The forecast chart for deep-water wind waves (see page 131) shows how ocean wind waves grow higher and longer as the speed, duration, and fetch of the wind increase. In this way, a 20-knot wind that has blown over a fetch of 200 nautical miles is seen to have a wave height of 8 feet, while a 35-knot wind will produce a 25-foot wave if the fetch is 400 miles.

TABLE 15. CLASSIFICATION OF WAVES	
h/L	Terminology
$0 \rightarrow 1/20$ $1/20 \rightarrow 1/2$ $1/2 +$	Shallow-water waves Intermediate-depth waves Deep-water waves

h = depth of water   L = wavelength

Waves behave very differently in shallow water (where the water depth is less than one-twentieth the wave length). When deep-water waves approach shallow water, their speed and height can increase significantly as the wave literally "feels" the bottom. If the water depth decreases gradually, the waves can also change direction. But if the bottom rises rapidly, the waves will be reflected the way light waves bounce off a mirror. Where the water gradually becomes shallow, a wave front approaching the shoreline at an angle will rotate in such a way that it will wash ashore nearly head-on. The cause is bottom friction. As the advancing portion of an angling wave front approaches shore, it is slowed down while the trailing sector continues to move at a faster speed. Soon the wave has been rotated, making its final approach nearly perpendicular to the shore. As the water becomes even more shallow, the wave steepens and speeds up until breakers form and the waves die out.

In shallow water the effect of fetch is virtually unimportant. Wave development depends only on the speed of the wind and the depth of the water, not on the size of the lake or bay or on the length of time the wind has blown. Shallow-water waves develop quickly to their full potential and have shorter wavelengths than their deep-water counterparts. Together, these two features— rapid buildup and short wavelength—explain why shallow waters present a greater hazard to the boater than deep waters.

If the height of the wave becomes more than one-seventh (15 percent) of the wavelength, the wave becomes unstable and the top of the crest breaks away from and moves ahead of the rest of the wave. Such waves are easily identified by whitecaps, a potentially dangerous situation for the boater. In the stable wave, the water mainly moves up and down in a cyclic fashion. The waves are not a flowing of the water, but rather a periodic up-down oscillation. This can be verified by observing how a flat, floating object (one not blown by the wind) does *not* move forward with the waves. But when the wind creates whitecaps, water is pushed forward of the crest and the object moves. The momentum of the water pushed ahead is enormous (computed as the product of the speed and weight of the breaking water) and can far exceed the momentum of a boat attempting to make headway.

What can be done to modify this effect? Some relief is possible when fish oil* is released very slowly (about a gallon per hour) from a boat that is hove-to with a sea anchor. The oil creates

---

*The viscous properties of petroleum make it much less effective than fish oils.

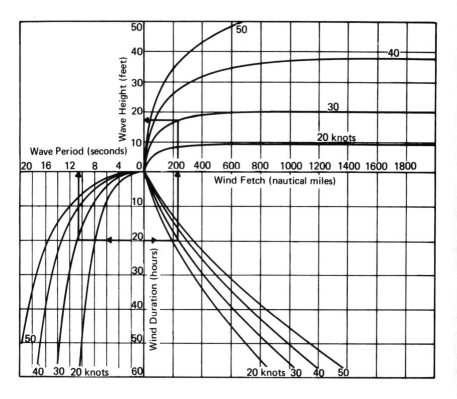

FORECAST CHART FOR DEEP-WATER WIND WAVES

1. Enter the lower right section with the speed and duration of the wind; where they cross, go upward to determine the wind fetch. Example: wind speed = 30 knots and duration = 20 hours; fetch = 230 nautical miles.

2. Enter the lower left section and compute the wave period the same way. Example: wave period = 11 seconds.

3. Enter the upper right section with fetch. Proceed vertically to intersection with wind speed; then go horizontally left to read the estimated significant wave height, that is, average height of the tallest third of all waves. Example: significant wave height = 17 feet.

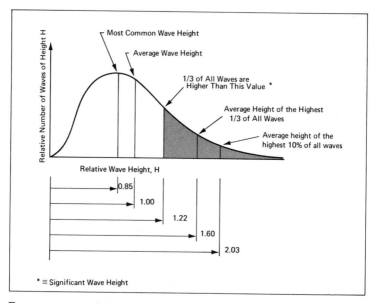

Deep-water wave-height spectrum. *At any time, wind waves in deep water have a wide variety of heights. There are a few very small waves, a small number of unusually tall waves, and the vast majority falling somewhere between the two extremes. If you measured the height of every wave for one or two hours and then made a graph of the number of waves that you observed having various heights, you would create a wave-height spectrum like the one above. If you assign the average wave height a relative value of 1 (for example, this could represent one fathom, or 6 feet), then you would find that the most common wave height was 15% shorter than the average. Oceanographers often talk about the "significant" wave height; strictly speaking, this is the average height of the tallest one-third of all waves present. Experience has shown that it is 60% taller than the average wave height.*

a very thin film on the water surface that increases the surface tension of the waves, while the slick surface also reduces the drag force that the wind is able to exert. The production of whitecaps will be inhibited within a 100-foot radius. This is not to say that the technique will work for all wind waves, but in an emergency the method can be quite helpful to the small-boat operator.

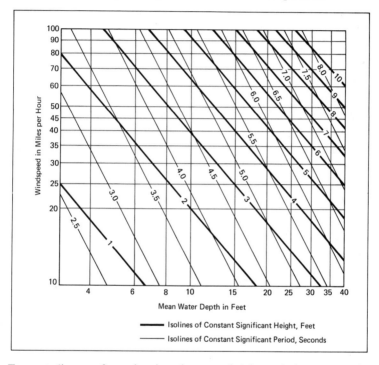

Forecast diagram for estimating the wave height and the wave period of wind-driven, shallow-water waves

Reading the Wind

Because wind is so important to the boater, it follows that he should know how to judge the average speed of the wind, what *peak wind* speeds to expect, and how to recognize the signs that foretell wind changes. Judging the actual (or true) wind from a moving boat is not easy, even with an anemometer aboard. The wind you feel or measure is an *apparent wind,* a combination of the true wind and the movement of the boat. The situation is further complicated because the wind continually fluctuates in speed and direction, even though the average wind may be unchanged. On top of these rapid fluctuations, there are changes in the average speed and direction. The vulnerability of the boater to wind makes it essential that he be on the lookout for signs in the sky or on the water that foretell wind changes.

True Versus Apparent Wind Wind is air in motion. Riveted in one location, we perceive air moving past us as wind. But we can think of wind another way. If the air all around us is totally calm and we are moving, then we also feel wind—the air appears to move, although in reality it is at rest.* When your motorboat cruises northward at 10 knots on a calm day, you feel a northerly wind (moving southward) blowing at those same 10 knots. Were there easterly winds also blowing at 10 knots, the apparent wind would come from the northeast and have a speed of 14 knots. In this second example, the 10-knot easterly wind is the true wind; the 10-knot northerly wind is the *boat wind,* and the 14-knot northeasterly wind is the apparent wind.

The Beaufort table (Table 4) found in Chapter 1 provides a guide for estimating the true wind speed on land and water by observing wind effects on waves, trees, and so forth. Having an

---

*Actually this is not totally correct. When the air is calm and some object moves through it, a certain amount of air will "cling" to the surface of the object and be dragged along. This type of air movement is called *drag flow.* Streamlined objects create little drag, while those that cause significant drag are sometimes called *bluff bodies.* An automobile is one example of a bluff body. As it moves down the highway, the air next to the surface of the car is dragged along at slightly more than half the car's speed. This explains why leaves on the hood of your car do not always blow away as you are moving: The relative wind on the hood of a vehicle moving at 20 mph is only a few miles per hour—not enough to sweep the hood clean. The hood is one portion of the car where the drag is greatest; other locations experience less wind resistance and have smaller drag. By streamlining their vehicles, automobile designers try to reduce the energy used by the car to pull the air along with it.

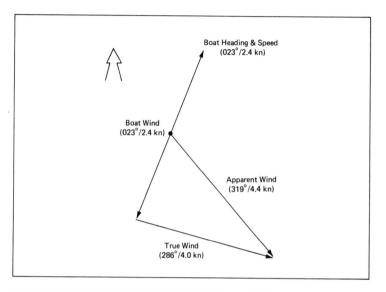

HOW TO DETERMINE THE TRUE WIND FROM A MOVING BOAT

1. Plot the heading and speed of the boat.
2. Plot the "boat wind" that is due solely to the movement of the boat (equal to the speed of the boat but with reverse heading).
3. From the origin of the boat vector, plot the direction and speed of the apparent wind as measured with the boat's anemometer and wind vane.
4. Draw a vector from the tip of the boat wind to the tip of the apparent wind; this is the true wind vector.

anemometer aboard is a more accurate technique, but it only measures the apparent wind speed, and you will need to account for the boat wind to determine the true wind. Wind direction can be estimated more easily. Pennants on the boat indicate the apparent wind direction, while the drift of smoke or the flapping of a shore-based flag will give a reliable indication of the true wind direction. Cloud and wave movements are less reliable. Waves typically blow at an angle of about 15° to the true wind direction, while even low-level clouds may move considerably differently from the wind where it affects you—at the surface.

Average Wind Condition and Gusts Suppose for the moment that the forecast of tomorrow's wind conditions calls for "southwesterly winds in the morning at 5 knots, increasing in the afternoon to 10 knots with gusts to 15 knots." What the forecast is saying is that averaging the morning's wind conditions will yield a southwesterly direction and a speed of 5 knots. Watching a wind vane and an anemometer continually would show you that the speeds fluctuated above and below the 5-knot average; almost never was the speed precisely 5 knots. Speeds of between 4 and 6 knots occurred quite frequently, while calms and 10-knot winds were less frequent. Wind speed fluctuation is unpredictable. Fluctuations may be rapid or slow, but usually the two occur in combination with the rapid fluctuations superimposed on the more gradual changes. During warm, sunny days the frequency of the fluctuations is greater than on overcast days; nighttime winds usually have less frequent changes.

The variation of wind speeds over the course of an hour can be critical to the small-craft skipper. While the average wind speed is important, the stronger wind speeds that occur sporadically throughout the hour are of overriding concern. These short-term, high-speed winds are called *gusts.* Alan Watts, author of *Wind and Sailing Boats,* gives rules of thumb for estimating wind speeds in two types of gusts: the mean or average gust, and the maximum gust. The *mean gust speed* is the average speed of the wind when it is increased substantially above the average wind speed. The *maximum gust speed* is simply the peak wind speed that occurs. Table 16 provides a way to estimate gusts depending on the average speed, time of day (that is, atmospheric stability—relationship between heat of air and water), and type of fetch (land or water).

The more significant, longer-term wind fluctuations typically last from 1 to 10 minutes. The higher the wind speed, the shorter

TABLE 16. RELATIONSHIPS BETWEEN AVERAGE WIND SPEED AND GUST SPEEDS

Time of Day	Daytime		Nighttime		Day or Night	
Type of Fetch	Over Land		Over Land		Over Open Waters	
Average Wind Speed (knots)	Mean Gust Speed (knots)	Maximum Gust Speed (knots)	Mean Gust Speed (knots)	Maximum Gust Speed (knots)	Mean Gust Speed (knots)	Maximum Gust Speed (knots)
7–10	14–16	14–20	10–15	13–19	10–15	10–15
11–16	17–25	21–32	16–24	21–30	17–24	17–24
17–21	26–32	30–38	25–31	31–38	26–32	26–32
22–27	33–41	40–49	33–40	40–49	32–41	32–41
28–33	42–50	45–53	42–50	48–56	42–50	42–50
34–40	51–60	54–64	51–60	58–68	51–60	51–60

SOURCE: Alan Watts, *Wind and Sailing Boats* (Devon, England: David & Charles Limited, 1973), p. 41.

the period of the gust. Furthermore, the wind speed and direction normally are coupled during gusts: As the wind speed increases, the wind direction generally veers, or shifts clockwise. Conversely, a slackening of the wind speed is usually accompanied by a *backing*, or counterclockwise shift in direction. Over open waters, these directional changes are usually on the order of 10°, while 20° changes are more typical of winds that have evolved over land (inland and coastal waters).

Signs of Wind Change There are many causes for a changing wind, and most of them can be anticipated accurately if you are on the lookout. Cumulus clouds are one common cause of wind changes that last on the order of minutes. First, be aware of the direction in which the clouds are moving; more often than not it will differ from that of the winds near the water (or land) surface. Air in front of the individual cumulus clouds is gently sucked in toward the cloud; this leads to a calming effect ahead of the cloud as the inflow toward the cloud opposes the large-scale flow of the air. For the larger cumulus clouds—cumulus congestus and cumulonimbus—this effect is quite pronounced and is commonly referred to as the calm before the storm. At the peak of maturity, there is a sudden thrust of colder air that flows down from the forward half of the cloud and pushes out ahead of the cloud. Wind speeds typically increase dramatically in this so-called gust front. The wind direction will also have a tendency to shift, aligning itself along the direction of movement of the cloud itself. At the trailing edge of the cloud, the wind speed falls off and its direction changes in the opposite way.

Wind changes associated with the passage of weather fronts are also quite predictable. The fronts themselves often can be recognized by the cloud sequence that precedes them (as in the case of warm fronts), by the clouds that mark their presence (particularly cold fronts), or by consulting a weather map. Ahead of a warm front, winds will usually be southwesterly, shifting to westerly or west-southwesterly after the frontal passage. Along the front the winds may be somewhat stronger, often decreasing after the warm front has passed. Cold fronts, on the other hand, are marked by significant changes in wind speed and direction. Before the front passes, ground-level winds typically are out of the southwest quadrant, but they quickly shift to northwest after the front has passed. Along the front, the winds frequently are quite strong and gusty and may remain so for long periods after the frontal passage.

Other predictable wind changes are those that take place from day to night. When the large-scale wind flow is not particularly strong, evening will produce a calming effect that is maintained until after sunrise. If the evening wind does not die down, there may be an approaching storm (although it can also mean that the strong pressure gradient that is to blame will bring yet stronger daytime winds, but no precipitation).

Perhaps the most ominous of all the wind signs the boater will ever face are those that indicate the approach of a hurricane.

Hurricanes

A man on Long Island had satisfied a lifelong ambition by ordering an expensive barometer from Abercrombie & Fitch. It arrived the morning of September 21st. Eagerly, he unwrapped it and was disgusted to find the needle stuck at "Hurricane." After shaking it in a vain attempt to start it working right, he sat down, wrote a very stuffy letter to A&F, and went right out to mail it. When he returned, his house was gone and the barometer with it.

—*The New Yorker*
November 12, 1938

Hurricanes are neither the largest nor the most intense storms that mariners are apt to encounter. Tornadoes are more intense and low-pressure storm systems much larger. But the combination of size (300 to 400 miles in diameter) and ferocity (winds of from 74 to 200 mph) make them the most destructive storms on earth. What then do you need to know about hurricanes? Three things!

One, learn to respect them. If you have not experienced firsthand the winds, the waves, and the surge, read on to get at least a feeling of what to expect. Two, know when and where hurricanes are likely to strike and where you can get accurate and up-to-date weather information day and night. And three, plan ahead. Know beforehand what actions may be necessary to protect yourself, your home, your boat. Yacht clubs and marinas should consider developing hurricane preparedness programs to ensure that all necessary equipment, information, and emergency plans are available if and when a hurricane strikes.

The 1938 hurricane that pounded the East Coast with 100 mph winds and 30-foot waves left a wake of destruction behind: 600 dead, 2,000 injured, and $500 million in property damage. And it was not unique. Galveston, Texas, was devastated on September 8, 1900, with a loss of 6,000 lives. More recently, Audrey (1957) took 500 lives and inflicted $200 million in property losses, while damages from Agnes (1972) were estimated at $3 billion. These four are only a sampling of the hurricanes that strike the United States at an average rate of three per year. But, by comparison, none has ever approached the almost unimaginable loss inflicted in 1970 when a hurricane out of the Bay of Bengal killed 300,000 persons in the coastal regions of Bangladesh.

Hurricanes occur in all the world's tropical oceans, except the South Atlantic. Virtually all United States hurricanes strike the Gulf and Atlantic coasts, although one will occasionally stray north of Baja to lash at southern California. The term "hurricane" comes from the Amerindian storm god Hurakan and is strictly reserved for Atlantic and South Pacific storms. In the western Pacific and the China Sea they are called *typhoons*.

Atlantic hurricanes are generally born anywhere within a belt that extends from the Gulf of Mexico, through the Caribbean Sea, to the tropical North Atlantic near Africa. An average of six Atlantic storms occur each year, although the number has ranged from two (1929) to twenty-one (1933). About two-thirds of all hurricanes that strike the United States hit during August and September; the remainder between June and November.

Portrait of a Hurricane For all of its power, the hurricane has a tranquil beginning. The late summer sun heats up large masses of warm, moist ocean air. Then, sometimes, a disturbance in the normal easterly wind-flow pattern causes the warm air in the middle to rise as cooler air streams in from the sides (a kind of "atmospheric chimney"). A large atmospheric whirlpool forms with air spiraling counterclockwise about the center in the Northern Hemisphere. The rising air cools to form raindrops and clouds, while releasing enormous amounts of heat energy. So great is this energy supply that if one day's heat could be totally converted to electricity, it would supply the electrical needs of the United States for six months.

As the winds spiral into the developing storm's center, a large centrifugal force is set up that maintains the open center or *eye* of the storm. The cloud-free eye is 10 to 20 miles in diameter and has low wind speeds (less than 15 mph). Temperatures in the eye are 10° to 15°F warmer than in the outer zones of the hurricane. The boundary of the eye is clearly marked by a towering wall of nimbostratus and cumulonimbus clouds that rise to heights of 40,000 to 50,000 feet. The clouds surround the eye in a series of spiral bands. The heaviest rains and strongest winds occur in a 60- to 70-mile band around the wall. Also, lines of thunderstorms (squall lines) frequently precede the hurricane by 300 to 400 miles.

At first, a hurricane moves its one million cubic miles of air slowly to the west at about 250 miles per day. Although hurricanes are highly unpredictable, they will usually curve toward the north and northeast where the Gulf and East coasts are often the victims. At this time, their speed may increase to 400 miles per day. In all, a hurricane may travel 1,800 miles over an average life span of nine days. Once it encounters land, the hurricane usually loses its force rapidly and takes on the appearance of a more typical low-pressure storm system. Although the rough land surface saps

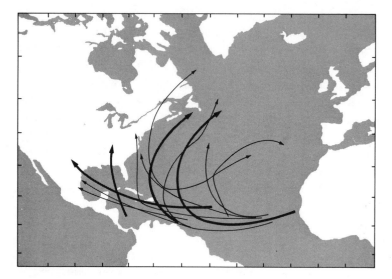

Principal origins and tracks of Atlantic hurricanes

some of the hurricane's energy, it is the loss of the heat supply from the warm ocean waters that brings about the hurricane's end.

Hurricane Weather Typical rainfalls from a passing hurricane range from 6 to 12 inches per day, or 10 to 20 billion tons of water. In 1950, Hurricane Easy lingered over Cedar Keys, Florida, for three days, depositing 24 inches of rainwater. The greatest hurricane deluge on record was 100 inches in a single day in the Philippines. As damaging as the resultant flooding often is, many areas depend on hurricanes for some or most of their water supply. For example, one-fourth of the annual rainfall in the southeastern states comes from hurricanes.

For winds to be called hurricane-force, they must register a minimum of 74 mph. Hurricane winds of up to 200 mph have been observed, while 100-mph winds are not uncommon. These winds generally blow within a 50-mile radius of the center, although gale-force winds (greater than 37 mph) are found as far out as 200 to 300 miles. Ocean waves produced by these winds may easily reach heights of 35 to 50 feet, and on occasion may be 100 feet in monster storms. Shipping and low-lying coastal communities are particularly susceptible to these waves, which can travel outward from the storm's center at a speed of 1,000 miles per day.

An unpleasant by-product of the hurricane is the spawning of tornadoes. Once thought to be a rarity, tornadoes are now believed to occur relatively often during hurricane passages over land. Fortunately, they are not as intense as those that form during spring and summer in the Midwest.

The perils of wind, rains, and waves cannot compare to the damage inflicted by the ocean *storm surge*. Often incorrectly called a tidal wave, the storm surge is the rapid rise of tide—perhaps

Hurricane damage in Biloxi, Mississippi. Courtesy of the National Oceanic and Atmospheric Administration

Breaking waves driven by hurricane-force winds. Courtesy of the National Oceanic and Atmospheric Administration

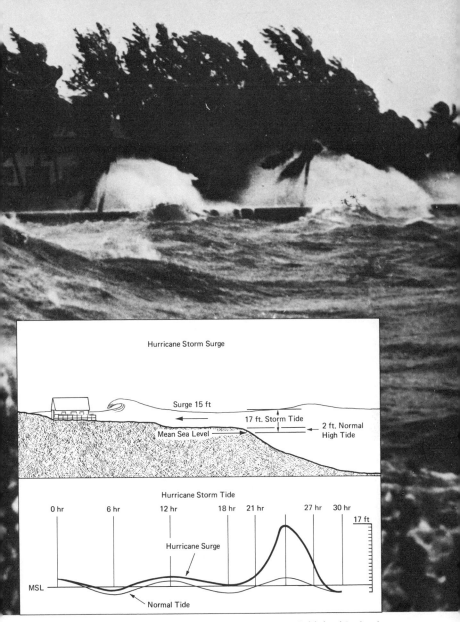

The combination of battering waves, storm surge, and high tide is the hurricane's most dangerous feature.

TABLE 17. HURRICANE-DAMAGE RATING SYSTEM

Category	1	2	3	4	5
Weather Conditions*					
Winds (mph) Surge (ft)	74–95 4–5	96–110 6–8	111–130 9–12	131–155 13–18	> 155 > 18
Potential Property Damage from High Winds†	Shrubbery (m) Trees (m) Unanchored mobile homes (m) Flimsy signs (m)	Foliage (s) Trees (m) Exposed mobile homes (m) Flimsy signs (s) Roofs, doors, windows (m)	Foliage (s) Trees (s) Flimsy signs (s) Roofs, doors, windows (m) Small buildings (m) Mobile homes (s)	Foliage (s) Trees (s) Most signs (s) Roofs, doors, windows (s) Mobile homes (s) Small buildings (m–s)	Foliage (s) Trees (s) All signs (s) Roofs, windows, doors (s) All mobile homes (s) Small buildings (s) Large buildings (m)
Potential Property Damage from Storm Surge	Low coastal roads inundated Minor pier damage Some small boats torn from exposed moorings	Coastal roads may be cut off 2–4 hours before arrival hurricane center Marinas flooded Small craft torn from exposed moorings	Serious coastal flooding Coastal buildings destroyed/ damaged by tides, waves, debris Low-lying escape routes cut 3–5	Coastal terrain below 10 feet mean sea level flooded to 6 miles inland Major damage to coastal buildings within 1,500 feet of shoreline Low-lying escape	Major damage/ destruction to all buildings within 1,500 feet of shoreline Escape routes cut off 3–5 hours before arrival hurricane center Major beach erosion

		hours before arrival of hurricane center Coastal terrain below 5 feet height flooded to 8 miles inland	routes cut off 3–5 hours before arrival of hurricane center Major beach erosion	Massive evacuation of low-lying areas to 10–15 miles of coast
Comments	Shoreline and low-lying areas to be evacuated	Further evacuation	More extensive evacuation	

*Only winds *or* surge need reach these levels, although both may.

m = moderate damage

s = severe damage

HURRICANE SAFETY RULES

1. ENTER EACH HURRICANE SEASON PREPARED. Every June through November, recheck your supply of boards, tools, batteries, nonperishable foods, and the other equipment you will need when a hurricane strikes your town.
2. WHEN YOU HEAR THE FIRST TROPICAL CYCLONE ADVISORY, listen for future messages; this will prepare you for a hurricane emergency well in advance of the issuance of watches and warnings.
3. WHEN YOUR AREA IS COVERED BY A HURRICANE WATCH, continue normal activities, but stay tuned to radio or television for all National Weather Service advisories. Remember, a hurricane watch means possible danger within 24 hours; if the danger materializes, a hurricane warning will be issued. Meanwhile, keep alert. Ignore rumors.
4. WHEN YOUR AREA RECEIVES A HURRICANE WARNING, PLAN YOUR TIME before the storm arrives and avoid the last-minute hurry which might leave you marooned, or unprepared.

   KEEP CALM until the emergency has ended.

   LEAVE LOW-LYING AREAS that may be swept by high tides or storm waves.

   LEAVE MOBILE HOMES for more substantial shelter. They are particularly vulnerable to overturning during strong winds. Damage can be minimized by securing mobile homes with heavy cables anchored in concrete footing.

   MOOR YOUR BOAT SECURELY before the storm arrives, or evacuate it to a designated safe area. When your boat is moored, leave it, and don't return once the wind and waves are up.

   BOARD UP WINDOWS or protect them with storm shutters or tape. Danger to small windows is mainly from wind-driven debris. Larger windows may be broken by wind pressure.

   SECURE OUTDOOR OBJECTS that might be blown away or uprooted. Garbage cans, garden tools, toys, signs, porch furniture, and a number of other harmless items become missiles of destruction in hurricane winds. Anchor them or store them inside before the storm strikes.

   STORE DRINKING WATER in clean bathtubs, jugs, bottles, and cooking utensils; your town's water supply may be contaminated by flooding or damaged by hurricane floods.

   CHECK YOUR BATTERY-POWERED EQUIPMENT. Your radio may be your only link with the world outside the hurricane, and emergency cooking facilities, lights, and flashlights will be essential if utilities are interrupted.

KEEP YOUR CAR FUELED. Service stations may be inoperable for several days after the storm strikes, due to flooding or interrupted electrical power.

STAY AT HOME, if it is sturdy and on high ground; if it is not, move to a designated shelter, and stay there until the storm is over.

REMAIN INDOORS DURING THE HURRICANE. Travel is extremely dangerous when winds and tides are whipping through your area.

MONITOR THE STORM'S POSITION through National Weather Service advisories.

BEWARE THE EYE OF THE HURRICANE

If the calm storm center passes directly overhead, there will be a lull in the wind lasting from a few minutes to half an hour or more. Stay in a safe place unless emergency repairs are absolutely necessary. But remember, at the other side of the eye, the winds rise very rapidly to hurricane force, and come from the opposite direction.

5. WHEN THE HURRICANE HAS PASSED:

SEEK NECESSARY MEDICAL CARE AT RED CROSS disaster stations or hospitals.

STAY OUT OF DISASTER AREAS. Unless you are qualified to help, your presence might hamper first-aid and rescue work.

DRIVE CAREFULLY along debris-filled streets. Roads may be undermined and may collapse under the weight of a car. Slides along cuts are also a hazard.

AVOID LOOSE OR DANGLING WIRES, and report them immediately to your power company or the nearest law enforcement officer.

REPORT BROKEN SEWER OR WATER MAINS to the water department.

PREVENT FIRES. Lowered water pressure may make fire fighting difficult.

CHECK REFRIGERATED FOOD for spoilage if power has been off during the storm.

REMEMBER THAT HURRICANES MOVING INLAND CAN CAUSE SEVERE FLOODING. STAY AWAY FROM RIVERBANKS AND STREAMS.

SOURCE: National Oceanic and Atmospheric Administration

as much as 15 feet or more. It is the result of both the low atmospheric pressure, which literally raises the level of the water, and the piling up of the water by the onrushing winds. The storm surge can flood low-lying coastal areas, causing catastrophic damage and loss of life as in the Bangladesh and Galveston disasters.

The National Weather Service has developed a system to rate the potential destruction from hurricanes. It is based on the maximum speed of the storm's winds *or* the surge, whichever is more dangerous. When reported, it is only to characterize present conditions in the storm and not what future conditions may be. Table 17, pages 146 to 147, summarizes the rating system, along with examples of the types of damage that can be anticipated.

Planning Ahead   Planning ahead is the only way to deal effectively with a hurricane. Yet few groups or individuals have developed hurricane-preparedness programs to:

- Acquire and distribute accurate and up-to-date hurricane wind and surge information
- Ascertain before the storm hits the extent and severity of damage that is likely in the area (according to the intensity of the storm)
- Take effective steps to strengthen or change moorings, pull boats from the water, shore-up doors and windows, and, if necessary, evacuate to higher ground

Although a few special items might need to be acquired, the main element in putting such a plan together is *forethought.* You should determine the rise in sea level in a harbor or marina that corresponds to each of the five storm categories. It is then easy to make a checklist of what properties will be damaged and what preventive steps can be taken to minimize damage in the twenty-four-hour warning period you will normally have before the storm surge and high winds strike.

To stay abreast of current weather conditions and forecasts, two continuous radio services are available in most coastal areas: National Weather Service VHF weather broadcasts (162.40 or 162.55 MHz) and the Transcribed Weather Broadcast Service of the Federal Aviation Administration (200–400 KHz or 108–118 MHz).

A few hours' planning now just might pay big dividends later.

Chapter Seven Mountains and Valleys

Mountains—even small ones—are weather makers. More than any-place else on earth, mountain-valley regions produce significant and, at times, life-threatening changes to the winds, temperature, precipitation, and lightning.

On the large scale, mountain ranges affect and modify the climate of an entire region either by blocking or modifying air masses. This explains, for example, why the climate of Florence, Italy, is so mild at 44°N latitude, while Munich, Germany (250 miles to the north), is cold, cloudy, and damp by comparison. Situated between these two historical cities are the Alps, which effectively block the movement of storm systems from northern Europe. The California coastal range modifies the onshore Pacific air to create a different climate in Sacramento from the upwind one in San Francisco. Further downwind, the Sierras continue to modify the maritime air until desert conditions prevail in Reno, Nevada. From San Francisco to Sacramento, yearly rainfall averages drop from 22 to 18 inches, while the percentage of daylight hours with sunshine increases from 65 to 74. At Reno, 150 miles inland, the yearly precipitation totals less than 8 inches.

Mountains and valleys affect weather on a more localized scale as well. Individual mountains can trigger summer showers or create howling winds through narrow passes. Together with valleys they produce complex wind and temperature patterns that vary both with location and hour of the day. On an even more localized scale, individual meadows, hollows, and canyons create temperature and wind patterns varied enough to be the difference between survival and disaster. Spending the night in a barren, snow-covered hollow might lead to frostbite during an emergency situation, whereas a campsite on a knoll under a stand of pine trees (using boughs underneath you as an insulator) will be far more tolerable.

TABLE 18. AN EXAMPLE OF THE INFLUENCE OF A MOUNTAIN RANGE ON REGIONAL WEATHER: TACOMA, WASHINGTON, AND YAKIMA, WASHINGTON			
	Geographical Location		Difference:
Weather Element	Windward Side of the Cascades— Tacoma, Wash. (elev. 172 ft.)	Leeward Side of the Cascades— Yakima, Wash. (elev. 1068 ft.)	Tacoma (windward) minus Yakima (leeward)
Temperatures:			
Winter average	40.7°F	30.6°F	+10.1°F
Summer average	62.4°F	68.8°F	− 6.4°F
Average 24-hr range	14.8°F	29.0°F	−14.2°F
Average 365-day range	24.7°F	44.5°F	−19.8°F
Average relative humidity	67%	46%	+21%
Average sunshine duration (for daylight hours)	43%	65%	−22%
Average number of clear days	73 days	113 days	−40 days
Average number of cloudy days	185 days	161 days	+24 days
Wind speed:			
Winter average	8.1 mph	5.9 mph	+ 2.2 mph
Summer average	7.6 mph	7.7 mph	− 0.1 mph
Average yearly precipitation	38.7 in.	7.2 in.	+31.5 in.

SOURCE: Helmut Landsberg, *Physical Climatology* (DuBois, Pa.: Gray Printing Co., Inc., 1962), p. 300.

Because mountains have such a great effect on the weather, you stand a good chance of forecasting conditions in the short term (even if they are apt to be severe on occasion). By knowing how the upper-level winds are blowing, the type and extent of clouds, the humidity, the time of day, and the local geography, you will be able to learn for yourself what weather changes to expect. You must be tuned in to the conditions around you and have a good memory (or notebook) for past weather happenings. So if you are backpacking in the Cascades, for example, and you observe persistent westerly winds above the influence of the mountains, prepare yourself for the fog and drizzle that may soon develop because of the transport of moist air from the ocean. In the Rockies, the meaning of a west wind would be different. To the west lie the intermountain deserts of Utah and Nevada that signal warm and dry weather for Colorado. Knowing what to expect also demands an awareness of what weather is typical of the season and the region. Thunderstorms are infrequent during the summer in the Cascades—fewer than ten days. But in the Colorado Rockies, nearly one in two summer days has a thunderstorm. Just as batters know what to expect from a pitcher they have hit against before, the outdoorsman must learn to read the local weather signs of the region he frequents.

Mountain weather can affect your physical well-being in many ways. As the elevation increases, so does the intensity of sunburning ultraviolet radiation. The sun's intensity is about 25 percent greater at 10,000 feet than it is at sea level. When puffy cumulus clouds dot the sky, their white surfaces act as floating mirrors to reflect additional ultraviolet radiation from the sun on your skin and in your eyes, and onto plants that also can be sensitive. Two other factors make the sunburn problem even more acute on mountain slopes. If the surface is sloped toward the sun, the energy it receives will be proportionately greater than for a flat and level surface; when the ground is snow covered, nearly all of the sun's radiation is reflected upward. The sunburn potential can be very great if all four factors are present at the same time: (1) high elevation, (2) scattered cumulus clouds, (3) southerly sloping surface, and (4) fresh snow cover. It's not hard to understand why skiers and climbers can receive bad burns on clear early-spring days.

Mountains affect you directly in other ways as well. As the elevation increases, the pressure and density of the air decrease—the air becomes thinner and its oxygen content decreases. There are no noticeable effects below 7,000 feet, but permanent inhabitants above 10,000 feet show signs of anemia and muscle weaknesses.

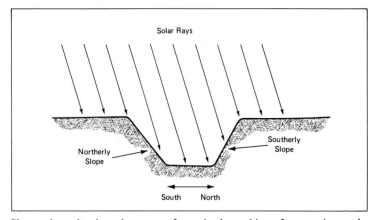

Slope orientation is an important factor in the making of mountain weather. *The southerly slope receives much more solar energy than a flat surface when the midday sun is to the south during winter, while the northerly slope receives even less.*

Some hikers experience dizziness and sleeplessness for the first day or two, and then these symptoms usually disappear. Even in the oxygen-poor air at 14,000 feet, some communities live in the Andes with no apparent inconvenience.

Profile of Mountain Weather Elements

Weather elements in mountainous regions vary from the upwind to the downwind side of individual mountains and mountain ranges as well as with elevation. In the remaining sections of this chapter, we will touch on the important weather phenomena you are likely to confront in the mountains or valleys, including temperature variations, winds, precipitation patterns and fog, thunderstorms and lightning, and flash floods. But before delving into these, you should have a good overview of just how variable weather can be in these regions. Some of the variations that occur between the windward and leeward sides of a mountain range have already been mentioned. But nothing much was said of the ways the various weather elements change as the elevation changes within mountainous regions. In his classic review, *The Climate Near the Ground,* Rudolf Geiger has presented an excellent synopsis of the typical changes that occur in climatic conditions with height above sea level. The information Geiger presents is taken from many studies of the Austrian Alps, but the picture the information paints is appropriate as a mountain-climate primer for other regions as well.

Table 19 (page 157) contains much of the information accumulated and discussed by Geiger. In scanning the table, bear in mind that these are average values that have been tabulated and they are strictly applicable to only one region. However, the trends reflected in the numbers are widely appropriate even if the absolute values and the more subtle implications cannot readily be transferred to other regions. The list highlights some of the key points:

- The sun's intensity increases steadily with elevation under both clear and overcast sky conditions. At 10,000 feet there is 20 percent more solar energy in the summer than is available near sea level when the skies are clear. Even when skies are overcast, there is still considerable solar energy at the 10,000-foot level; in June, the overcast value is about 60 percent of the sea level value for clear skies.
- Temperatures fluctuate more widely near sea level than they do higher up.

- The decrease in temperature with increasing elevation is nearly twice as great in summer as in winter. The average July vertical temperature gradient is precisely the same as the average value of the "free atmosphere" away from the ground (3.5°F decrease per 1,000-foot elevation increase); this is an indication of the presence of convection currents during summer. (Convection currents are updrafts that occur when the air is unstable, either due to marked warming of the ground surface or the presence of a cold air mass overlying a warmer one.) The milder winter gradient (about 2.2°F per 1,000 feet) reflects a lesser exchange of air between the mountain sides and the free atmosphere, and a greater influence of the mountain surface on temperatures.

- Relative humidity is significantly higher on the crest than in the valley, thereby making it more difficult to stay warm in winter (most clothing is a poorer insulator at high humidities) and cool in summer (as the rate of body-cooling evaporation is slowed down).

- Precipitation increases dramatically with elevation; more than three times more precipitation falls at 7,200 feet than at 650 feet. In winter, snowfalls are more frequent and deeper as elevation increases. And the time when the maximum snow depth occurs becomes progressively later in the year as the elevation increases; for every 2,000-foot rise in elevation, the time of maximum snow depth is another month later and the snow is considerably deeper.

A tabulation such as Table 19 can be an invaluable aid for planning an expedition or hike, although of course the data should be appropriate to the region you will be traversing. Sometimes the information may be available so that you can construct your own table, but you may have to collect your own wind, temperature, and humidity data. Appendix D lists a few simple weather instruments that are portable and should do the job.

TABLE 19. CHANGES IN CLIMATE WITH ELEVATION (ABOVE SEA LEVEL) OBSERVED IN AUSTRIAN ALPS

Elevation (feet)	Average Daily Total Solar Energy (calories per sq. cm) Clear Days June	Clear Days Dec.	Overcast Days June	Overcast Days Dec.	Average Air Temperature (°F) July	Jan.	Year	Yearly Number of: Summer Days	Frost-free Days	Days with Frost	Relative Humidity	Yearly Precipitation (inches)	Yearly Number of Days with: Dry Ground	Snow Cover	Snowfall Information: No. of Days per Year with Snowfall	Average Snowfall (inches per day)	Yearly Snowfall (inches)	Max. Depth of Snow: Depth (inches)	Date
650	691	130	155	30	67.1	29.5	48.2	48	272	93	71	24.2	187	38	27	1.8	20	8	Jan. 18
1300	708	136	168	32	64.9	27.5	46.4	42	267	98	74	29.5	173	55	32	2.0	46	12	Jan. 23
1950	723	141	180	34	62.8	25.7	44.8	37	250	115	77	34.8	160	81	38	2.3	72	20	Jan. 28
2600	735	146	192	36	60.8	25.0	43.5	31	234	131	78	40.4	147	109	45	2.5	97	29	Feb. 3
3250	747	150	205	38	58.6	25.0	42.3	15	226	139	76	45.7	133	127	53	2.8	123	37	Feb. 11
3950	759	154	220	40	56.5	25.0	40.8	11	218	147	74	51.0	120	138	62	3.0	149	39	Feb. 14
4600	771	157	236	43	54.2	24.6	39.2	7	211	154	73	56.3	107	152	73	3.2	175	47	Feb. 21
5250	782	160	253	47	52.2	23.2	37.0	4	203	162	73	61.8	93	169	85	3.5	201	56	Mar. 3
5900	791	163	272	50	49.8	21.0	34.9	2	190	175	74	66.9	80	189	98	3.7	226	66	Mar. 14
6550	799	166	293	54	47.7	19.2	32.7	0	178	187	74	72.2	67	212	113	4.0	252	78	Mar. 26
7200	807	168	314	58	45.0	17.2	30.6	0	163	202	75	77.6	53	239	128	—	278	95	Apr. 8
7850	814	169	336	62	42.6	15.4	28.4	0	146	219	78	—	40	270	143	—	—	117	Apr. 20
8500	821	170	358	66	40.3	13.5	26.1	0	125	240	80	—	27	301	158	—	—	144	May 3
9150	828	171	380	70	37.8	11.7	23.9	0	101	264	82	—	13	332	173	—	—	176	May 15
9800	834	171	403	75	35.2	9.7	21.7	0	71	294	84	—	0	354	188	—	—	215	May 29

SOURCE: Rudolf Geiger, *The Climate Near the Ground* (Cambridge, Mass.: Harvard University Press, 1965), pp. 444 and 445.

Temperature Variations

Maintaining body heat is essential to survival. As the air becomes colder, you will require more insulating clothing or your body will need to supply more energy—or both. Even if the situation is not critical to your survival, it can be uncomfortable. By surveying your environment and using some weather know-how, you can avoid unpleasant cold spots.

Localized Cold Spots The night is dry and clear, the ground is barren, and you are seeking a campsite on a plateau or in a meadow or valley. Does it make a difference how you choose the site? Maybe. If the air is calm, there may be cold pockets you should avoid. For now we are *not* referring to temperature differences due to variations in the makeup of the ground surface or soil type (these are also important and were discussed in Chapter 1). Instead, there can be cold spots created by a very gentle, almost imperceptible slope of the ground. While these cold spots are quite shallow (less than a few feet), they are sufficiently deep to envelop the reclining body of a camper in a sleeping bag. This is what happens: Clear, dry, and calm air favors cooling of the ground surface by radiation. As the air becomes colder, it also becomes denser or heavier and will gently flow downhill at a barely detectable speed (less than 2 mph). The slopes needed for the air to flow are gentle indeed. A surface that slopes 1° (about 18 inches in 100 feet) is adequate. As the air flows downslope, it pools in depressions (although the depressions are not an essential part of the phenomenon). Where the ground is higher, the downflowing air is replaced by a gentle drift of air from above. This air is warmer than the air originally at the surface. The result of this very localized air circulation pattern is that the air on knolls, mounds, or other "high" places is considerably warmer than the air in the lower spots. Temperature differences as large as 8°F have been measured between two points about 110 yards apart with elevation differences of only 5 feet.

These cold spots do not form until well after sunset when the cooling through radiation has overwhelmed the heat being supplied by conduction upward through the ground. The air motion is so slight that the person trapped in this river of cold air cannot sense its movement and therefore may not realize that warmth lies nearby and uphill. The cold spots will not form if a perceptible wind persists through the night, as the turbulent overturning of the air will continually bring down relatively warm air from above. The key to avoiding the cold "pools" is knowing when and where they are apt to occur, and then favoring relatively high ground for your campsite.

Thermal Belts Just as a gently sloping surface can produce cold pools of air, the larger-scale sloping sides of mountain valleys also set up unusual temperature distributions. The mechanism is some-what different, however, because the mountain valley situation is accompanied by appreciable wind circulations (these upslope and downslope winds are discussed later in this chapter). The result is that the cold nighttime air found during clear nights both on the valley floor and on the adjacent hilltops or plateaus is divided by a warm zone part way up the valley sides. The warm zone is often called a *thermal belt* and has important effects on the location of dwellings, orchards, vineyards, campsites, and much more. To better understand thermal belts and their significance, let's take a closer look at how, when, and where they form.

Most mountain valleys have a shape that resembles a broad wedge, very much like the one on page 154 used to illustrate the effect of slope orientation on the amount of solar energy received. During clear nights, the slopes and the valley floor and upper plateau cool off rapidly because of the loss of heat by infrared radiation. The cool air layer on the plateau gets thicker as the night and the radiation cooling progress. Some of this cold air slips off the edges and flows downslope, mixing with the cooling air along the slopes that is also sinking downhill. By comparison with the cold pools, the air moves quite perceptibly at speeds of above 5 mph. The movement causes turbulence in the cold airstream, much like the overturning of water as it flows downhill or over a dam. The cold, descending air thus mixes with the warm air that spans the gap across and above the valley floor (this reservoir of warm air is left over from the daytime heating within the valley; it is high enough above the valley floor and slopes so that it is not cooled appreciably at night). The air along the slopes is warmed in this way, as well as by the compression warming of descending air that raises temperatures by 5.5°F for every 1,000-foot drop in altitude. On the valley floor the air continues to be cooled by the escaping infrared radiation, although the side regions are modified by an influx of relatively warm (descending) air.

There are a number of factors required to create nighttime thermal belts: (1) clear, dry skies, (2) near-calm, large-scale winds, and (3) dry ground with sparse vegetation. The intensity and frequency with which these factors occur determine the characteristics of the thermal belt that forms. The dry, sparsely vegetated slopes of the Pomona Valley in California have resulted in thermal belts where the nighttime minimum temperature was 28°F warmer than the valley floor 225 feet below. While this may be an extreme example, thermal belts typically have minimum temperatures that range from 8° to 16°F higher than in the valleys. Thermal belts do not span the entire slope. Rather, they usually are only a few hundred feet thick. Temperatures are not uniformly warm throughout the belt; they are highest in the middle and taper off on both the upper and lower sides. The height of the zone above the valley floor is not uniformly predictable, but instead varies with the shape and depth of the valley. Thermal belts do tend to form on the lower half of the slope. While each valley is unique, thermal belts most generally seem to lie in a band from 200 to 1,200 feet above the valley floor.

Now that you have an idea of what happens, you may be able to spot thermal belts by keeping an eye out for some telltale signs. In Europe, many older villages, estates, and monasteries were built within these belts. Trees bloom several days earlier and some plants emerge from the soil up to twenty days earlier in this belt than in the valley below. In the spring, the snow melts here first. There are other locally indigenous signs as well that you will come to recognize with experience.

Elevation Effects The two preceding sections focused on some very specialized and important aspects of temperature variations with altitude. During daylight hours, conditions are quite different from the nighttime ones that produce thermal belts.

On the average, temperatures decrease nearly 3°F per 1,000 feet of elevation rise. Mountains force the air upward on the windward side. When winds are strong and the upflow is pronounced, the temperature drop may reach 5.5°F per 1,000 feet. If the upflow results in the formation of clouds, heat is released in the condensation process and the rate of temperature drop is decreased by about 1.5°F per 1,000 feet.

Temperature variations up the mountain depend also on the nature of the air mass, in addition to time of day and orientation

The Cloister at Ettal in the Bavarian Alps was built within the thermal belt to obtain the most favorable climate in the mountain valley.

of the slope. The graph below gives some representative profiles of temperature that are appropriate to spring conditions and both continental and maritime air masses. The absolute temperatures are different for individual days and seasons, but the overall nature of the temperature variations with elevation is similar.

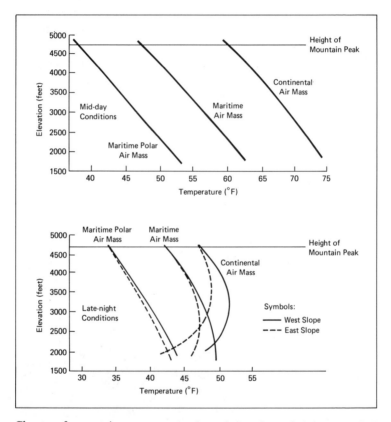

Change of mountain temperatures depends on time of day, type of air mass, and orientation of the slope. *Shown here are springtime conditions in the Alps.* Adapted by permission from Rudolf Geiger, *The Climate Near the Ground* (Cambridge: Harvard University Press, 1965), copyright 1950, © 1957, 1965 by the President and Fellows of Harvard College.

Winds

Isolated hills and mountains as well as mountain-valley systems create their own wind circulations and modify existing wind patterns. New wind circulations are established in response to the heating and cooling patterns that are set up by the terrain; these are called *thermal circulations.* Where the terrain modifies existing wind patterns, the winds are said to be "forced." Forced winds are found when the large-scale wind flow is strong. Thermal circulations occur on sunny days when the large-scale wind flow is weak, and on clear nights when the lower layers are decoupled from the upper-level winds. Often, however, the actual wind conditions are a combination of thermal and forced effects.

Thermal Circulations   There are two principal types of thermal wind circulations in mountainous terrain: *thermal slope winds* and *mountain-valley winds.* Thermal slope winds are found along the slopes of isolated hills and mountains, ridges, ravines and canyons, and mountain valleys. As their name implies, mountain-valley winds do not form near isolated hills or along broadly sloping terrain. Instead, they form only in terrain where broad and deep, well-defined valleys are found.

The temperature pattern that leads to the formation of thermal belts also causes a drainage of cold air downhill at night and an ascent of warm air uphill during the day. The *upslope winds* begin fifteen to forty-five minutes after sunrise and reach peak speeds of 4 to 8 mph at solar noon or, in the case of partly cloudy skies, when the solar energy at the ground is greatest. Southerly slopes receive the largest amounts of energy and therefore also have the strongest upslope winds. Northerly slopes may experience almost no upslope winds. Upslope winds avoid exposed ridges and promontories and favor gullies and ravines. Even on southerly slopes the uphill flow of air can be quite fickle. Floating cumulus clouds that shade the sun can turn the upslope winds off and on like a light switch. These daytime winds tend to be intermittent. When they gust, extreme speeds may reach 20 to 25 mph. Friction from the ground surface reduces the low-level winds. Peak speeds are usually found about 100 to 200 feet above the sloping surface, while the entire layer of upslope air is 300 to 600 feet thick for the larger mountains. The upslope motion often is intense enough to form a line of small cumulus clouds along ridges.

By comparison, *downslope winds* are gentler, steadier, and confined to a more shallow layer. They start blowing fifteen to forty-five minutes after sunset and continue until sunrise. The same set of weather conditions that promotes the development of thermal belts favors the nighttime downslope breezes.

The *glacier wind* (also called a *fall wind)* is a special type of downslope wind. As its name signifies, this wind is found over glaciers. It has the unique characteristic of blowing downslope both night *and* day because the ice surface is nearly always colder than the overlying air. Sometimes a glacier may not cover the entire slope. Then a daytime downslope wind over the glacier is common, as is a warm upslope wind further downhill. Where the cold and warm flows meet, the cold air slips beneath the ascending warm air and continues downhill until it loses its cold identity by mixing with the warmer air. Depending on the size of the glacier, fall winds may exist in a layer that is as shallow as 150 feet or as thick as 1,400 feet.

When the sloping surfaces form a semienclosed valley, the wind pattern is different. Not only are there upslope and downslope winds, but there are *upvalley* and *downvalley winds* as well. Seen from above, these valleys are open on one end and resemble the letter V or U. The open end often blends into a flat plain, while the narrow end often has higher elevations. The winds form in much the same way as the slope winds, except that they occur on a larger scale because of the larger dimensions of the valley axis in comparison with the sloping sides. The larger scale makes the valley winds stronger than the slope winds, but they form more slowly because of the valley's gentler slopes; once set in motion, however, the valley winds' greater momentum keeps them blowing longer. It can be compared to pushing a heavy car: It takes longer to start than a wagon, but it will roll for a greater time. The upvalley winds blow during the daytime along the valley axis from the wider, lower end toward the narrower, upper end; downvalley winds blow at night in the opposite direction.

For most of the day the valley winds do not exist by themselves. They form a complicated, three-dimensional wind pattern as they interact with the slope winds. The eight-part illustration on page 166 shows the interplay of the slope and valley winds throughout a daily cycle. The meteorological conditions most favorable for the development of valley winds are identical with those required for slope winds: light, large-scale wind flow, strong daytime heating, and pronounced nocturnal cooling due to radiation heat loss at night.

Upslope winds dominate the wind pattern (see diagram) in mid-morning (1) when the temperature rise is sharpest in the valley and along the slopes; the valley winds are calm at this time while they are reversing from the nighttime downvalley flow in anticipation of the upvalley motion that will begin shortly. The mid-morning air circulates across the valley in a closed pattern: up the sides, across the valley at ridge level, and down toward the center of the valley floor. By midday (2), the upvalley winds have set in; they strengthen the upslope motion along the slopes and reinforce the downdrafts over the valley center. The strong updrafts produce a line of cumulus clouds above the ridges (when the air is sufficiently moist), while the downdrafts keep the valley center cloud-free. The upslope winds die out in late afternoon (3), but the upvalley winds persist. The cumulus clouds dissipate as the sustaining upslope winds cease blowing. The downslope winds begin their downhill flow after sunset (4); however, the wind pattern is complicated by the persistence of the day's upvalley winds, which coexist and interact with the downslope motions. It is only in

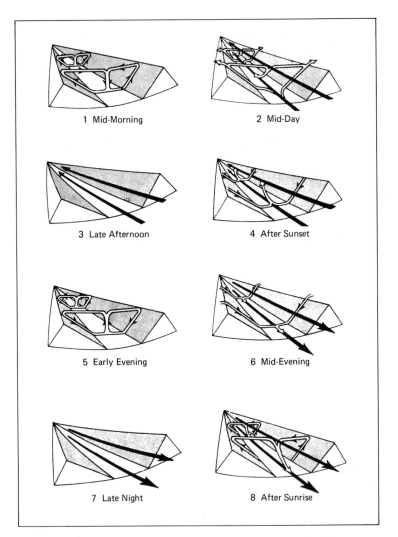

1 Mid-Morning

2 Mid-Day

3 Late Afternoon

4 After Sunset

5 Early Evening

6 Mid-Evening

7 Late Night

8 After Sunrise

The daily cycle of interplay between valley and slope winds. Adapted with permission of the American Meteorological Society, from Compendium of Meteorology (1951), p. 665.

the early evening (5) that the upvalley winds stop, but it is still too early for the night's downvalley flow to begin. The winds during early evening form a closed cycle across the valley that is a mirror image of the midmorning situation. Now the cold air flows down the slopes and across the valley floor where it encounters the drainage winds from the opposite slope. Here in the middle of the valley the air is gently pushed upward, then flows outward at ridge height to complete the cycle. A few hours later (6), the downvalley winds form and interplay with the downslope winds. Later in the night (7), the downvalley winds wash out the downslope winds and dominate the wind picture in the valley. It is only after sunrise (8) that the upslope winds begin to counter the downvalley flow.

The complexity of the valley wind pattern is apparent. The wind you feel at any instant and at any particular location may bear little resemblance to the general pattern described here. Slopes are not uniform, and neither are the slope and valley winds they produce. The interplay among the various wind forces is not simple, and neither is the transition throughout the day from one wind regime to another. Invisible battles among the various wind forces occur constantly, and the resulting winds are frequently gusty and intermittent.

Forced Circulations   Some aspects of forced wind circulations were discussed in Chapter 6. To quickly summarize what was said there: Wind flow over and around *individual* hills and mountains is affected by the stability of the air. Unstable air prefers to blow over the obstruction, while stable air mainly goes around the object. When the air is neutral, winds tend to be diverted both over and around. However, mountainous terrain usually consists of a series of hills, valleys, mountains, and passes, rather than isolated peaks. This means that many types of mountain winds occur as a result of the way the air is channeled through passes and into valleys, and deflected over ridges and crests down into the lowlands that lie below. When the existing, large-scale winds encounter mountainous terrain, a number of changes occur: Winds become stronger and more gusty, and there may be significant effects on the air temperature and the occurrence of cloud formations. Note that the forced wind circulations discussed here produce the temperature changes, whereas the thermal circulations described earlier are caused by temperature changes.

Whenever the wind encounters any type of topographic feature, there will be changes in the speed and direction of the airflow. It would be impossible to describe all of the different possibilities that exist in nature. There are, however, some very basic features you should recognize and anticipate. By knowing these fundamentals and using your own experiences in the mountains, it will be quite possible to understand the winds you encounter and anticipate the winds you may face at some future time or location.

Whenever there is a marked variation in the landscape and the ambient winds are sufficiently strong, a portion of the airflow will detach itself from the general wind pattern and create a wind *eddy*. An eddy is simply a localized wind pattern that is significantly different from the general windflow through the area. The sea breeze and slope winds discussed earlier are eddy circulations. So are the vortices that form at the trailing edge of an airplane wing. Eddies also form wherever the wind encounters an object that presents a (partial) barrier to its flow. If the "object" is a steep cliff, eddies will form regardless of whether the wind blows into or away from the cliff. Eddies will not form, however, if the object is streamlined with smooth, gentle changes in height or breadth. Eddies can form almost anywhere: upwind, downwind, or above sharp cliffs, ridges, or buttes. Wherever eddies form, wind speeds fluctuate widely and rapidly, giving rise to gusty conditions. Instead of providing shelter from the wind, the region behind steep slopes often will have peak wind speeds greater than in the open areas farther away. The wind direction near the ground within an eddy is opposite to the large-scale wind flow. In most cases, the eddies are transient features. They form near the variations in the terrain,

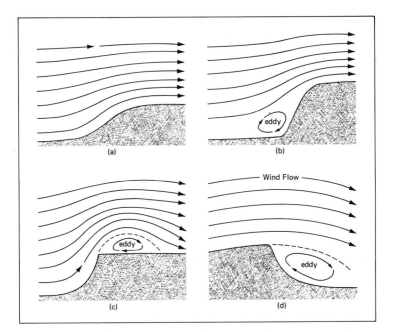

Eddies form in the wind flow near steep slopes. *Eddies do not form when the terrain is smooth or rolls gently (a); instead, the wind flow is nearly parallel with the ground. An eddy will form when the wind encounters a steep slope (b); at the base of the slope the wind direction reverses and blows opposite to the prevailing winds. Sometimes the upwind slope is streamlined, thereby preventing formation of an eddy. Instead, the slope enhances the uphill movement of air. An eddy will also form on top of a hill or butte where the transition from the sloping side to the flat top is abrupt (c). Eddy formation on the downwind side (d) is a mirror image of the pattern on the upwind slope.*

but after a short time detach themselves from their "birthplace." They then move downwind in the airstream, dissipating as they go. As soon as one eddy has become detached and moves away, another forms and the cycle repeats itself. Only in special circumstances when the air is stable and the winds are light and very steady will the eddy remain permanently attached.

Even when the wind flow over a terrain feature does not create an eddy, there is still an appreciable effect on the wind speed. (See diagram.) The wind speed accelerates on the upwind slope as the ascending air is compressed between the terrain and the undisturbed airflow higher up (D). On the downwind slope, the speed decreases from its peak on the crest (B) to the base (C). Because energy is taken out of the air by ground friction, the wind on the lee side (C) is slower than its windward counterpart (A). During winter, the accelerating air on the windward slope picks up and transports snow, which is then deposited on the leeward slope where the air moves more slowly. With time, the transfer of snow from windward to leeward deforms the symmetrical shape of a hillock or ridge. The deformation on the leeward side, in turn, can lead to eddy circulations and further changes in the distribution of windblown snow. *Snow cornices* are formed in this way along ridgecrests. Coastal sand dunes share many of these same features, except that the lower surface tension of the sand inhibits the intricate formations that are so typical of the more exotic snow cornices.

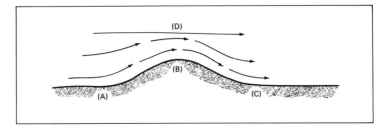

Wind flow over a terrain feature

When the wind flow is nearly head-on to a lengthy range of hills or mountains, a unique type of wind pattern can result in the downwind region. The pattern is characterized by a train of *lee waves* (also called *mountain waves*) in the airflow. Lee waves provide soaring pilots with the updrafts they need to remain aloft, but the turbulence that often accompanies them can present a distinct hazard to all light aircraft. On the ground, lee waves are an important part of mountain weather because of the very strong and gusty winds they create. Because our emphasis is on weather on and near the ground, I will not stress the aerial aspects of lee waves; the interested reader is referred to more definitive books on weather and soaring.*

Picture a stream of water flowing over a submerged log lying across the streambed. Downstream of the log, the surface of the water undulates because of a series of waves having characteristic wavelengths and amplitudes. The waves do not move with the flowing water stream; they are stationary below the obstruction. Air also behaves like a fluid, and when an airstream encounters a line of hills or mountains, a similar wave pattern forms. The features of these lee waves depend on the characteristics of both the airstream and the mountains. Soaring pilots in the Sierras have ascended to altitudes of 45,000 feet by taking advantage of the great height and strong upward air motion of mountain lee waves. In extreme cases, lee waves up to heights of 60 to 100,000 feet have been observed. Closer to earth, lee waves commonly create surface winds with average speeds of 50 mph and gusts to 125 mph. Boulder, Colorado, is particularly vulnerable to these strong winds. Roofs are frequently torn off houses, power lines snapped, windows shattered, and automobiles sandblasted. One wind storm in the late 1960s caused more than $2.5 million in damages to property in Boulder.

*An authoritative and comprehensive coverage of weather and soaring is given in the *Handbook of Meteorological Forecasting for Soaring Flight,* Technical Note No. 158, Secretariat of the World Meteorological Organization, Geneva, Switzerland (1978).

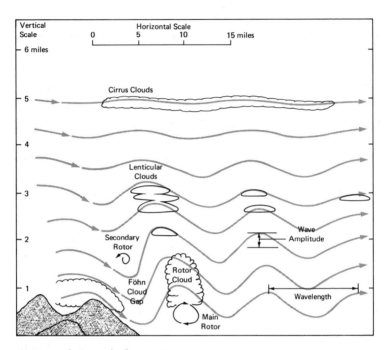

Features of mountain/lee waves

Four meteorological conditions must be satisfied for the formation of lee waves: (1) The large-scale winds must approach the ridgeline at an angle less than 30°; (2) the change in the wind direction with altitude must be small—less than 20°; (3) the air must be stable at the height of the ridgecrest; and (4) wind speeds must be strong—more than 18 mph for hills less than 3,000 feet high, and more than 30 mph for large mountains of around 13,000 feet. A number of topographical features are also important: the height and shape of the mountains, and the spacing between successive ridgecrests. The wind speed is the major determinant of the wavelength of the lee waves. When one or more wavelengths equal the spacing between successive ridges, the lee waves are amplified by the upslope motion induced by the terrain—the lee waves and terrain features are said to be *in phase.* But when the terrain spacing is not an exact multiple of the wavelength (the two are *out of phase*), the lee waves are reduced or destroyed over the downwind ridgecrest. In the case of Boulder, there are no mountains farther downwind (eastward) and the lee waves are neither depressed nor amplified by the terrain.

Although lee waves are infrequent in the San Francisco Bay Area, they do occur. The axis of the Bay is oriented northwest-southeast. On parallel lines run the East Bay hills (elevation about 2,000 feet) and the coastal hills (similar elevation) on the west. Large-scale windflow from the northeast is rare; southwesterly flow, however, will sometimes occur. The marine inversion along the central Pacific Coast usually has its base around 1,500 feet—a good sign for lee waves since it provides a stable air layer aloft. Because the hills are so low, wind speeds need only exceed 18 mph. Since southwest winds come off the smooth ocean, wind speeds in this range are quite common. What then is the effect of the successive lines of hills that are separated by 20 miles? A rule of thumb gives the following relationship between wind speed and wavelength:

Wind Speed (mph)	Wavelength (miles)
23	1.86
35	3.72
46	5.59

For this example, assume that the wind speed is 35 mph; the wavelength of the lee waves would therefore be 3.72 miles. The crest of the fifth wave northeast of the coastal hills would lie 18.6 miles downwind and the crest of the sixth, 22.3 miles. This would lead to the elimination of significant updrafts over the East Bay hills because the lee waves are out of phase with the terrain. At the location of the inland hills, the lee wave is in its trough and has a strong downward motion. The hills create an opposing upslope motion, and the combination of the two effects leads to

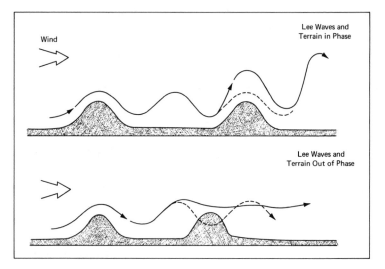

Wind

Lee Waves and
Terrain in Phase

Lee Waves and
Terrain Out of Phase

Wind

Topography can modify lee waves

nearly horizontal airflow over the hills. With a slower wind speed, say 23 mph, the two effects are in phase and the upward motion over the East Bay hills would be amplified.

Occasionally, if conditions are just right, an eddy phenomenon called a *rotor* will occur. Rotors are stationary eddies that form when the amplitude of the lee waves is particularly large; for example, in the lee of tall mountains, or when terrain spacing is in phase with the lee wave spacing. The strongest rotors are found under the crest of the first wave; with a 25-mph wind, about 2 miles downwind of the ridgecrest. Rotors feature strong updrafts on their windward side. This strong vertical motion produces a *rotor cloud* when the air is sufficiently moist. Although the rotor cloud does not move from its position just above the actual rotor, air is continuously moving into, through, and out of it; the cloud is actually in a constant state of simultaneous development and destruction as fresh air circulates through it. The rotor cloud has its base at the height of the mountain top, while the rotor itself is contained below the mountain top. Surface wind speeds near the rotor are intense and gusty. Recordings indicate that the wind speed can increase from 10 to 100 mph in a matter of seconds.

The greatest frequency of strong lee winds along the east slope of the Rockies is during winter. There is a pronounced peak in January, although strong winds are common from November through March. Thirty-six percent of the high-wind cases occur between midnight and 6 A.M., another 27 percent occur between

Lenticular clouds formed by lee waves over Madison, Wisconsin. Courtesy of Leonard F. Hall

6 P.M. and midnight, while the remaining 37 percent are spread more or less uniformly throughout the period from 6 A.M. to 6 P.M.

Mountain valleys and passes can produce equally dramatic effects on the wind flow. As the large-scale winds blow across mountain valleys at an appreciable angle, the lee wave pattern breaks down. In its place, there is a strong tendency for the air in the valley to blow along the axis of the valley. Much the same phenomenon occurs in the deep street canyons of high-rise cities like New York and Chicago: The winds at ground level nearly always blow parallel to every street because of terrain channeling. The strong street-level wind speeds are, in turn, due to the constrictions in the airflow created by the various buildings. With less space available to it, the wind must speed up to keep the volume of air moving through the city equal to the volume of air approaching the city

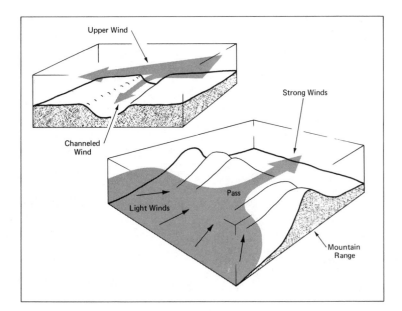

Mountain valleys and passes modify the direction and speed of the wind.

on its upwind boundaries. Mountain passes do the same. While some air is carried over the crests, a large amount is forced through the passes, creating nearly permanent high-wind regions.

The *chinook* is a very special type of lee wind. Not only is it strong and gusty, but it brings with it very large and rapid increases in air temperatures. Chinook is the common name used in North America; it is an Indian word meaning "snow eater." In Europe and throughout much of the world these hot, dry, gusty winds are called *foehn* winds or, more properly, *der Föhn*. It has other local names wherever it occurs: In the Andes it is the *zonda;* in Java, *koembang;* in Yugoslavia, *Ijuka.*

Technically speaking, the foehn is the result of a mountain lee wave and a few additional factors that cause the large temperature rise on the downwind plains. The temperature rise can occur as the result of one or both of two effects. All foehns are warmed by atmospheric compression at the rate of 5.4°F for each 1,000 feet of elevation drop as the air flows down the lee side of the mountain. From the ridgecrest of a 6,000-foot mountain, the air will warm 27°F by the time it reaches a 1,000-foot-high plain below. Yet this is no warmer than the air at the 1,000-foot level on the windward side of the mountain barrier—*unless,* that is, clouds form in the ascending, cooling air on the windward slope. With the appearance of clouds, heat is released to the air as condensation takes place and the rate of cooling with altitude is decreased. The exact decrease in the cooling rate depends on the moisture content and pressure of the air, but it is about 2° to 2.5°F per 1,000 feet. Going back to the example, the appearance of clouds halfway up the 5,000-foot windward slope would indicate that the temperature drop from the base to the top of the mountain is only 21°F (starting at the base, the air cools at the rate of 5.4°F per 1,000 feet for the initial 2,500 feet; thereafter in the condensing air, the cooling rate decreases to about 3°F per 1,000 feet). This means that the air would be only about 6°F warmer on the downwind plain than on the upwind plain.

The condensation-compression effect explains the dryness of the air on the leeward side and its ability to *evaporate* snow covers quickly on the downwind plains. In extreme cases, the dry foehn can remove up to 2 feet of snow in a single day! But the condensation process *cannot* explain the large, rapid rise in temperature that usually is experienced on the leeward plains. One extreme event saw a foehn wind increase the temperature by 49°F in two minutes in Spearfish, South Dakota. While the more normal temperature rise with the foehn is 15° to 20°F, larger increases are not all that rare. During the more dramatic foehns, the enormous temperature increase is mainly the effect of a rapid change in air mass; the condensation-compression only adds fuel to the fire. Both the Alps and the Rockies act as barriers that separate warm, moist maritime air on the windward side of the mountains from the cold continental air to leeward. When conditions are right for the formation of lee waves, the already warm maritime air rises up the windward slopes, forming clouds and depositing precipitation. The air then warms by compression as it descends the downwind slopes, where it shoves back and replaces the cold continental air. The abrupt change of conditions at the boundary of these dissimilar air masses can also lead to the development of low-pressure storm systems.

TABLE 20. SOME EXAMPLES OF TEMPERATURE RISE DURING FOEHN (CHINOOK) EPISODES IN RAPID CITY, NORTH DAKOTA

Date	Temperatures (°F) Before	After	Temperature Rise (°F)	Time Interval of Temperature Rise
Jan. 12, 1911	13°	43°	30°	10 minutes
Jan. 13, 1913	−17°	47°	64°	14 hours
Dec. 28–29, 1933	0°	67°	67°	18 hours
Jan. 22, 1943	−5°	54°	59°	4 hours
Feb. 23, 1957	−3°	59°	62°	12 hours

Foehn episodes last from a few hours to several days. Slightly more than half persist for one day or less; about one-fourth last one to two days; and about one in seven continue more than two days. Foehns are most frequent during the spring months in the Alps, during winter in the Rockies. Apart from the strong winds, extreme warming, and dryness (relative humidities may only be 10 to 20 percent), the foehn is frequently recognized by its lee wave-type cloud formations and excellent visibilities. Lens-shaped *lenticular* clouds frequently form at the crests of the lee waves, while a wall of flat stratocumulus clouds called the *foehn wall* typically sits immobile on the top of the ridgecrest.

In addition to the pronounced weather change it brings, the foehn also stimulates a wide range of physical ailments in a sizable fraction of the population. Some people suffer from high blood pressure during a foehn, and many European physicians avoid performing eye surgery during these episodes. Because foehns are so common in the Alps, front-page headlines frequently report them and the numbers of people suffering from headaches, depression, and high blood pressure.

Local, gusty wind circulations abound in mountainous areas. While each is important locally, it would be a virtually endless task to attempt to describe all of them. Variations of the chinook wind, for example, include the *Newhall winds* and *Santa Ana* winds in California. Two related, though different, wind circulations are worthy of mention before taking up lightning effects in the mountains. These are the *bora* and the *mistral.* The bora is a cold downslope wind. As with the foehn, warming occurs by compression as the air flows downslope. However, with the bora, warming is overshadowed by the initial coldness of the deep air mass. At the base of the downwind slope, the still-cold air replaces relatively warm air, creating a sharp drop in temperature and often-violent winds. The most famous bora occurs on the coast of the Adriatic Sea when cold, continental air migrates out of Russia, through Hungary, and spills down the low but steep mountain slopes along the Yugoslavian coast. The *mistral* is a special type of bora. Winds are usually stronger in the mistral because the air is speeded up as it is forced through constrictions in the topography. The mistral occurs along the western Mediterranean coast of France; a similar phenomenon along the northern Pacific coast of North America is called a *norther.*

Precipitation and Fog

Clouds form and precipitation follows when two criteria are met: The air must be moist, and there must be an upward transport of air to lower temperatures. Updrafts are abundant in mountainous terrain because of the thermal and forced wind circulations previously described. As a result, precipitation is frequent and heavy in the mountains whenever the resident air mass is sufficiently moist. Precipitation amounts normally increase with elevation because of the updrafts, which continue to cool the air as it rises. Mountains create as many downdraft regions as updraft when forced circulations prevail. As a result, the leeward sides of mountain ranges and individual peaks have much less precipitation than the windward sides.

The increased amount of precipitation on mountains is particularly remarkable in the moist tropics. The southwesterly Indian monsoon brings warm, moist air out of the Bay of Bengal up and over the Himalayan foothills where the famous weather station at Cherrapunji, India, records an average in excess of 33 feet of precipitation every year. Even more extreme are the windward slopes of Waimea Canyon and Kawaikini Peak on the Hawaiian island of Kauai, which receive an annual rainfall of nearly 42 feet.

Fog patterns are more subtle than precipitation. Valleys are nearly always free of fog when it rains or snows (unless the cloud bases reach down to the valley floor). The likelihood of fog during precipitation episodes increases dramatically with altitude. Typically,

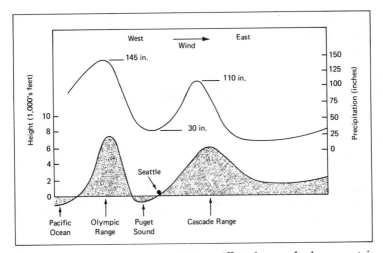

Precipitation amounts and patterns are affected severely by mountain ranges.

three days in four that have precipitation also have fog where the altitude is more than 2,500 feet above the level of the upwind plains or the valley floor. The exact altitude varies from place to place, but the trend is always toward more frequent fogs in the higher elevation during periods with rain or snow. The reason, of course, is quite simple: Fog basically is a cloud that touches the ground; and the higher you go, the closer you get to the base of the cloud deck.

When it is not raining (or snowing), fog formation tends to follow the same prerequisites that lead to the thermal belt: light winds and clear skies. Where the air is coldest, the likelihood of fog is greatest. Under these conditions, fogs form at night in the valley and on the slopes and plateaus *above* the thermal belt. When a thick nighttime fog forms, it can act as a barrier to the sun's energy and thereby reduce daytime temperatures. This can intensify the next evening's fog or cause it to form sooner. If the same fog-favoring weather pattern remains long enough, the fog in the valley may persist throughout the daytime period as well.

Flash Floods

The increased frequency and intensity of precipitation in the mountains can result in the very acute danger of flash floods. Flash floods can strike almost without warning. Campers near small streams have been swept away and drowned by flash floods on days when it didn't even rain at their campsite.

What is a flash flood? It is a sudden, short burst of water. Its forward edge will often resemble a virtual wall of water 10 to 20 feet high moving at speeds of up to 25 mph. It usually occurs as the result of intense rainfall over a very localized area. Only severe thunderstorms are capable of producing such short and heavy rainfalls. The great speed and height of the flood result from the channeling of the rainwater into and through a deep and narrow streambed or gorge. The intensity of the rainfall prevents the ground from absorbing much of the water. Frozen ground and rocky outcrops restrict the retention of rainwater and can also be contributors to flash floods, but are not essential. Ice jams and dam failures are two additional causes of flash floods.

Regions susceptible to flash flooding should be avoided. Previous flash flood areas sometimes can be recognized by the debris left behind at the high-water locations. Local residents can be helpful in defining dangerous areas. Be particularly careful when thunderstorm activity is intense, and avoid the lower stretches of long gulleys and streambeds where the accumulation of water is greatest. Since you will have only seconds or minutes for escape, know the flash flood safety rules ahead of time and select possible escape routes before you need to think of using them.

FLASH FLOOD SAFETY RULES

BEFORE THE FLOOD know the elevation of your property in relation to nearby streams and other waterways. Investigate the flood history of your area and how man-made changes may affect future flooding. Make advance plans of what you will do and where you will go in a flash flood emergency.

WHEN A FLASH FLOOD WATCH IS ISSUED listen to area radio and television stations for possible Flash Flood Warnings and reports of flooding in progress from the National Weather Service and public safety agencies. Be prepared to move out of danger at a moment's notice. If you are on the road, watch for flooding at highway dips, bridges, and low areas due to heavy rain not observable to you, but which may be indicated by thunder and lightning.

WHEN A FLASH FLOOD WARNING IS ISSUED for your area act quickly to save yourself. You may have only seconds:

1. Get out of areas subject to flooding. Avoid already flooded areas.
2. Do not attempt to cross a flowing stream on foot where water is above your knees.
3. If driving, know the depth of water in a dip before crossing. The road may not be intact under the water. If the vehicle stalls, abandon it immediately and seek higher ground—rapidly rising water may engulf the vehicle and its occupants and sweep them away.
4. Be especially cautious at night when it is harder to recognize flood dangers.
5. When you are out of immediate danger, tune in area radio or television stations for additional information as conditions change and new reports are received.

AFTER THE FLASH FLOOD WATCH OR WARNING IS CANCELED stay tuned to radio or television for follow-up information. Flash flooding may have ended, but general flooding may come later in headwater streams and major rivers.

SOURCE: National Oceanic and Atmospheric Administration

Flash flood damage. Courtesy of the National Oceanic and Atmospheric Administration

Lightning

Mountaineers, backpackers, campers, and hikers are all particularly susceptible to the hazards of lightning in the mountains. First, the ascending winds created by the rugged terrain cause thunderstorms to be more violent and more frequent. The mountainous portions of southern Colorado and northern New Mexico average up to seventy-five thunderstorms every year, while the adjacent plains states average between forty and fifty. Mountains are hazardous for a second reason as well: The rocky prominences that attract climbers and hikers also attract lightning discharges.

Lightning injures people when the body acts as a conductor of electrical energy. The brain and heart muscle are particularly sensitive to lightning damage, although other muscles and nerves are also susceptible. Injury or death commonly results from brain damage, heart failure or fibrillation, third-degree burns, or shock and subsequent injury from a fall. The flow of electrical current through the body will follow the path of least resistance (often the shortest path). With both hands touching a conducting pathway, the electrical current will flow through the chest and likely damage the heart or other vital internal organs. If the top of the head is in contact with or in close proximity to the pathway and the feet are on the ground, the current flow is through the head and down the spine. This suggests some positions that can minimize potentially fatal injury to the torso and head. One is to crouch with only the feet touching the ground—the current flow will be contained below the waist. Another is to sit down with feet and seat making ground contact, but not the hands or arms. Assuming these positions while riding out an electrical storm may not prevent serious injuries, but it may minimize them. Before worrying about how to stand or sit, it is imperative that you survey the area and determine where you should go to best protect yourself.

Before considering what locations are and are not safe, we need to explore the various ways that electrical current from a lightning strike can flow through your body. There are two principal ways this can happen. The first is a *direct lightning strike;* the second is by the current that flows through the ground as the result of a nearby lightning strike. The *ground currents* are less severe than a direct strike, but they are the more common lightning hazard and can cause significant harm due to deep burns and tissue damage.

There are a number of do's and don't's that can help you to avoid or minimize injury from direct lightning strikes:

- Seek protection near isolated prominences.
- — The top of the prominence should be at least 30 to 50 feet above your head.
- — The protected zone around the prominence lies within a 45° cone; keep your horizontal distance from the top of the prominence equal to or less than the height of the prominence above you.
- Maintain a horizontal spacing of at least 5 feet away from any vertical wall or cliff.

- Avoid exposed peaks and ridges and flat open terrain.
- Do not hold on to or stand near any vertical pole or rod.
- Do not stand beneath exceptionally tall or isolated trees; the entire canopy of the tree can pass an electrical current to the ground below (or through you, should you be there).

The ground current is more difficult to avoid. It is transmitted over the surface of the mountain along routes that make good conductors, such as crevices, lichen patches, mineral veins, firm and wet rock surfaces, and cracks filled with soil or moisture. The ground current is most intense at the point of the lightning strike and then falls off as the separation increases. The ground current not only moves along the mountain surface, but may jump a small depression or overhang, or follow a subterranean path into a cavern or cave. With these features of ground current flow in mind, some do's and don't's are apparent:

- Avoid crevices, lichen patches, cracks, and wet surfaces that are good conductors.
- Avoid overhangs, shallow depressions, and caves.
- If forced to seek shelter in a cave, avoid standing in the opening, near conducting paths, and within 5 feet of all vertical walls.
- Sit on a loose rock (preferably amid other loose rocks) away from vertical walls, depressions, and overhangs but within the 45° cone of protection from direct strokes.
- Insulate yourself from the ground or rock you are sitting on, using a dry blanket, sleeping bag, climbing rope, or similar item.
- Tie yourself down in precarious perches (across rather than along any conducting paths you can't avoid) in the event you should become unconscious and fall.

The best protection is not to be on the mountain during an electrical storm. Know how to determine when these storms will occur and then retreat to lower, safer elevations. The lightning avoidance tips may seem simple and easy enough to follow. However, shelter from lightning may be hard to find, and if you do locate a safe position, it will be cold, wet, and windy for the hour or more it takes to weather the storm. Not much fun, especially when you think a good forecast might have you spending the day around a warm fire in the valley below.

Chapter Eight **Forests**

Forests can affect weather both locally and more generally. They block the wind and the sun; they also prevent avalanches and reduce the probability and severity of damaging floods. When large forested regions are destroyed, the climate may be changed drastically. The state of Rajasthan in northwestern India may be just such a place. Nearly two thousand years ago, the region was widely forested and precipitation was abundant from the monsoon winds that blow out of the Arabian Sea into the area that is now Pakistan and India. With time the forests were harvested without being replanted. Even native grasses were unable to survive as a result of widespread overgrazing. What had been forests became wasteland. The wind attacked the soil, eroding it and filling the air with fine dust particles up to great heights. In turn, it was the dust—meteorologists presume—that sealed the transformation of the land from forest to desert. With large amounts of wind-driven dust suspended in the air, a significant fraction of the sun's energy was intercepted by the atmospheric dust layer. At the upper reaches of the dust layer, the air became warmer than it otherwise would have been over the forested terrain. Likewise, the interception of the solar energy by the dust layer made the surface cooler than before. Warmer temperatures aloft signify a more stable atmosphere and the suppression of the convection currents that aid the development of thunderstorms. More dust, less convection, reduced rainfall. Today the Rajasthan, Thar, or Great Indian Desert, as it is variously called, occupies about 100,000 square miles and may be earth's greatest monument to the consequences of indiscriminate deforestation.

Forests can affect the world's climate in yet another way. The worldwide average level of carbon dioxide in the air is partly controlled by the conversion of CO_2 to oxygen by trees and other plant life. With significant deforestation, carbon dioxide levels would increase, accompanied by a corresponding increase in the average global temperature owing to a strengthening of the greenhouse effect (see page 49). A warmer atmosphere would change growing patterns, ocean levels, atmospheric humidity, and perhaps cloudiness, winds, and the distribution of weather. All of these consequences may never take place, but they are theoretically possible. The rapid deforestation of broad sections of the tropical forests of the Amazon Valley has prompted some scientists to examine the likely effects on global climate.

The outdoorsman may be inclined to ponder the global consequences of deforestation while encamped in a stand of majestic conifers, but the immediate, localized weather effects are the ones with which he must cope. Virtually all the weather elements are altered by or in the forest: temperature, sunlight, wind, humidity, and precipitation. Not only does the forest affect weather, but meteorological conditions affect how the forest grows—and burns!

Weather Elements Within the Forest

Weather elements within an extensively forested area are significantly different from conditions above the forest, in the air overlaying adjacent open terrain, in small clearings within the forest, and on the edges of forests and near shelterbelts or windbreaks. Before examining the peculiarities of weather elements in clearings and discussing edge effects, let's start by looking at the microclimate the outdoorsman encounters within the forest—for argument's sake, at least one-quarter mile from clearings and the forest boundary.

Sunlight   Perhaps the first effect noticed in the forest is the light: There is less of it for shorter periods; there is also a certain amount of discoloration. These effects are most pronounced on clear, sunny days when there is little scattering of the sun's light in the air above the treetops. Most of the sun's rays strike the forest crown rather directly, like raindrops falling on a calm day. On the other hand, the light reaching the crown on cloudy and hazy days has been scattered by the airborne water and dust particles. Maintaining the analogy with precipitation, the scattered light rays strike the crown like windblown mist at virtually all angles. The scattered light rays can penetrate more easily through the forest crown, while the more direct light cannot.

The reduction in the light intensity is dramatic on clear days. Tropical rain forests reduce ground-level sunlight to less than 1 percent of the intensity above the crown. Forests in temperate latitudes can have relative light levels of only 2 to 3 percent. The exact amount of light penetrating the forest depends on the types of trees, their density, and their age. The usual trend is for young forests to let in less light than older ones; whereas a 10-year-old stand of firs may let only 10 percent of the ambient light through, a 100-year-old stand may transmit 25 percent. The effects of the percentage of the ambient light that is transmitted down through the forest canopy are not only sensed by the eyes. They are also felt in the temperature and humidity of the air, and are reflected in the growth on the forest floor. If there is no growth on the floor, the amount of transmitted light is less than 16 percent. Between 16 and 18 percent, undemanding mosses first appear. Berry plants will not grow until the percentage reaches 22 to 26 percent, while naturally seeded fir trees require 30 percent of the ambient light to penetrate.

The forest also affects color perception in coniferous forests and deciduous forests with foliage. We saw earlier that the light intensity is significantly reduced on the forest floor. But the reduction also varies accordingly to the wavelength of the light. Blues and purples are reduced the most. The green wavelengths are diminished in their intensity to about the same degree as the overall light intensity, while the reds are diminished the least.

The length of the daylight period is reduced considerably when the sky is cloudless. The time of civil twilight (roughly when a newspaper no longer can be read by natural light) is advanced up to sixteen minutes in a deciduous forest, up to twenty minutes in a coniferous forest, and up to twenty-eight minutes for an old, tall stand of firs or redwoods. Overcast and rainy conditions are even more extreme: Overcast skies can advance civil twilight forty-five minutes, while it may come nearly an hour earlier during rainy periods.

TABLE 21. LIGHT INTENSITY ON THE FOREST FLOOR COMPARED WITH LIGHT ABOVE THE CANOPY		
Tree Type (mature trees)	Relative Light Intensity (%)	
	Without Foliage	With Foliage
Red beech	26–66	2–40
Oak	43–69	3–35
Ash	39–80	8–60
Birch	—	20–30
Silver fir	—	2–20
Spruce	—	4–40
Pine	—	22–40

Temperature The effect forests have on temperature depends on many factors: time of day, thickness and type of the forest, cloud cover, and season. Forest temperatures differ most from those in the neighboring, open countryside on clear summer days and nights; the thicker the forest, the greater the effect. The forest canopy consists of the crowns of the individual trees, and affects both the incoming sunlight and the way the forest floor gives off longwave, infrared heat. Average air temperatures for a typical summer day (twenty-four hours) in the tree crown area are considerably warmer than those on the forest floor; the actual difference varies between 5° and 10°F from one location to another. The treetop area gets so much warmer because individual trees with foliage absorb about 95 percent of the sun's energy; the remaining 5 percent is reflected upward, while the open space between the trees permits a fraction of the total available sunlight to reach the ground. Not only is the average daily temperature warmer at crown level, but more important to the outdoorsman is the fact that the day-to-night change of temperature is considerably less at ground level inside the forest than it is either in the forest crown *or* at ground level in an adjacent clearing.

Daytime temperatures at the forest floor are cool, while crown temperatures are warm. At night, the situation is reversed: The air near the ground cools off only slightly and is relatively warm, while treetop air temperatures are considerably cooler. This means that the daylight period has warm air overlying cool air—a stable atmosphere condition that inhibits ascending air motions and causes daytime campfires to produce a smoky haze within the forest as the rising motion of the smoke is suppressed by the temperature inversion. The opposite situation prevails during clear nights. The air is unstable and updrafts are amplified because of the temperature decrease from ground to treetop. With sufficient openings in the canopy, the smoke escapes the tree zone only to become entrapped

TABLE 22. SOME TYPICAL DAY-NIGHT TEMPERATURE DIFFERENCES DURING A CLEAR AND WARM SUMMER DAY		
	Diurnal Temperature Range	
Location	Inside the Forest	Outside the Forest
Air, treetop height	33°F	—
Air, 6 feet above ground	24°F	47°F
Soil surface	19°F	57°F
Soil, 10 inches deep	2°F	6°F

TABLE 23. AVERAGE DIFFERENCE IN DAY-NIGHT TEMPERATURE RANGES BETWEEN AIR NEAR THE FOREST FLOOR AND AIR OVER OPEN, UNFORESTED TERRAIN

	Type of Tree		
Month	Beech	Fir	Pine
January	1.6°F	3.3°F	1.9°F
February	1.6°	4.1°	2.2°
March	1.3°	4.7°	2.4°
April	1.0°	5.1°	2.9°
May	4.5°	5.8°	3.6°
June	7.6°	6.2°	4.5°
July	8.1°	6.8°	5.2°
August	7.7°	7.0°	5.4°
September	6.8°	6.5°	5.3°
October	3.6°	4.6°	3.6°
November	1.6°	3.2°	2.0°
December	1.4°	2.8°	1.6°

Example: How much larger is the day-night range of air temperature in a beech stand during the months of April and July? During April, the average difference between high and low temperatures over open land is about 1°F greater than in the beech forest, but during July the temperature range is about 8°F greater over open land.

in the temperature inversion that now exists *above* the top of the forest.

Wind   Next to the reduced sunlight and the cooler daytime, summer temperatures, the wind (or lack of it) is one of the most noticeable weather features inside the forest. Wind conditions below the forest canopy bear little or no resemblance to conditions above or outside of coniferous forests or foliated deciduous forests. Wind speeds are light, and the direction is extremely variable. The lie of the land, the density of the forest, and the distribution of clearings and the like can all affect the light wind flow between ground level and crown height. The random drifting of smoke from campfires is visual evidence of the uncertainty with which the wind is directed.

TABLE 24. THE GENERAL NATURE OF THE VARIATION OF WIND SPEED WITH HEIGHT IN FORESTS*		
Height Above the Ground Relative to Tree Height	Relative Speed of the Wind	
	Inside the Forest	Over Open Terrain
120% (above the trees)	100%	100%
100% (treetop height)	80%	97%
95%	60%	96%
90%	40%	95%
80% (below the treetops)	20%	92%
55%	10%	85%
20%	8%	66%

With trees 50 feet tall and a 20-mph wind at 60 feet, the wind speed at 40 feet (80% of treetop height) is approximately 4 mph (20% of the undisturbed speed above the trees). By comparison, the wind under similar conditions over flat terrain would be about 18 mph at a height of 40 feet.

*Actual conditions will vary from this idealized picture.

Humidity   Relative humidities typically are 5 to 10 percent higher within the forest than in adjacent open terrain. The effect is most pronounced in the daytime. Many factors contribute: (1) The porous soil and the many roots take up and retain the water better than unforested land, (2) the trees themselves give off water during the daylight hours in a process called *transpiration,* (3) the lower daytime temperatures increase the relative humidity, and (4) the forest canopy and light winds inhibit the mixing of the humid air with the windier, warmer, and drier air above the trees and outside the forest. Nighttime conditions are different: Transpiration ceases, vertical mixing improves, and nighttime temperatures fall very little. Although the nighttime relative humidity does increase somewhat as temperatures slide, the increase is small in comparison with areas not shaded by the forest canopy. There, heat loss from the ground by infrared radiation is significant. Temperatures dip sharply, and dew or frost form more readily. You don't need to be in the forest to observe these effects; they can be seen in the vicinity of isolated trees everywhere. Autumn is a good time to see how trees prevent temperatures from dropping to the dew (or frost) point. With the trees still in leaf, clear and calm nights often result in temperatures falling to the frost point. A noticeable frost-free ring extends up to 6 feet around the perimeter of the tree. The sharp contrast between the still-green grass beneath the tree and the whitened frost-covered grasses around the rim of it makes the effect very striking.

When dew or frost forms on barren ground, it may also form on the top of the tree itself. Light dew forms only on the top of the outer layers of leaves, needles, and branches, but the entire tree can become coated when the dew is heavy. Campers below the tree then find themselves in a situation where it is apparently raining from clear skies as the dew on the trees forms large droplets that fall to the ground. The effect can perplex and annoy the unsuspecting camper, but the actual amount of dew that forms is quite small in absolute terms—certainly no more than .01 or .02 inch of water. Compared with the amount that forms in broad clearings, the dew on trees is significantly less—about half of its bare-ground counterpart.

Precipitation and Fog Forest campsites near the ocean or a large lake can be particularly vulnerable to significant amounts of precipitation during fair weather conditions devoid of rain-producing nimbostratus or cumulonimbus clouds. How? Forests act like large and very efficient sieves that sift the small water droplets out of fog. The droplets collect on the needles, leaves, and branches, where they grow by combining with other droplets. The resulting drops grow to sizes much larger than those that fall out of typical rain-clouds. Because of the large surface area comprised by the leaves and branches, forests collect appreciable amounts of precipitation: Some forests get up to 50 percent of their ground moisture this way. By comparison, tall grasses collect only about one-tenth as much water. Heavy fogs are known to produce nearly a quarter-inch of precipitation over a twelve-hour period. One extreme case on record resulted in more than *6 inches* of forest-filtered precipitation over a four-day period, while a nearby unforested section recorded not quite three-fourths of one inch of precipitation. Because the forests are so efficient in filtering out the fog droplets, there are great differences between the windward and downwind portions in the amount of precipitation that is obtained from the filtering process.

Fogs can severely damage the forests and injure campers when temperatures fall below freezing. Then the still-liquid droplets freeze on contact with the trees, forming *rime ice* on the needles and branches. Enormous weights build up rapidly and can cause branches and trunks to splinter or entire trees to topple. The same problem exists when high-tension power lines are exposed; weights of up to 30 pounds of ice per foot of cable length are known to occur.

The forest has equally significant effects on the amount of precipitation that reaches the ground and the way it is distributed— spatially, around individual trees, and temporally, around the clock. Without foliage, deciduous trees have nearly no effect on precipitation amounts or distribution. With foliage, there is a strong differential between the small amounts that reach the ground around the trunk (about 55 percent of the rainfall rate in the open) and the peak that falls around the rim or periphery of the tree (about 75 percent). But not all of the precipitation that falls on the tree reaches the ground, either during the rainy period or later. The amount "held back" by the tree is called the *interception;* it varies widely from 6 to 93 percent of the precipitation total. The intercepted precipitation clings to the tree and leaf surfaces and is then returned to the air by evaporation. The actual interception depends on the type of tree, the spacing between trees, the amount of foliage on deciduous trees, the speed and gustiness of the wind, and the character of the precipitation (light or heavy; small droplets or large drops). Interception is greatest with light rain and least during thundershowers; the forest will intercept about 25 percent of rain falling steadily at the rate of 1 to 1½ inches per day, whereas a similar amount falling during a three-hour downpour will result in almost no interception—about 2 percent.

The trees create a precipitation cycle of their own that bears little resemblance to the outpouring from the clouds. In the early stages of a light rain, the ground stays dry as the drops are intercepted by the leaves, twigs, needles, and other small tree members. This first stage may last for several hours, or until the crown has intercepted up to .1 inch of rain. The air at crown height normally has less than 100 percent relative humidity during the early stages, and so the intercepted moisture evaporates even while additional rain falls. With continued rainfall, the crown is unable to intercept and retain all of the drops that fall. Water then drips from the branches and trickles down the trunk. The drops that fall range from the size of the natural raindrops to a few that are quite large. "Natural" droplets typically have diameters less than .1 inch; oak trees "grow" drops that measure nearly a quarter-inch in diameter; silver fir are slightly smaller, and larch somewhat smaller yet at three-sixteenths of an inch. Of course, it is the rare outdoorsman who wants to know the size of the raindrops falling on his head. But the fact remains that the interesting pattern of drips and drops on your tent fly can be explained by interception and the large and therefore relatively infrequent drops that are produced. Lastly, interception causes the rainfall on the forest floor to persist for up to two hours after the cloud precipitation has ceased. During this period, *forest smoke* can often be spotted just above the tree crowns.

Forest smoke is the name given to the small, wispy patches of cloud and fog that form after the cloud rain has stopped; they usually persist for up to an hour, providing a picturesque transition after the dark and gloomy rainy period. The air in the tree crown is close to saturation (100 percent humidity) because of evaporation of the intercepted water. Occasionally a gust of wind transports a pocket of air upward, out of the crown and into the cooler air above. There the moist air from the crown cools a few degrees, reaches saturation, and forms a raggedy cloud.

Forests have equally pronounced effects on snowfall. Falling snow is initially snared by the branches, but with additional snowfall or increasing wind, the snow falls to the ground. The forest canopy then shields the snow on the ground from the sun's energy, minimizing the loss by evaporation. The lower wind speeds also slow down evaporation of the snow. As a result, snow loss in the forest occurs at a pace that is 20 to 30 percent slower than in the open.

Forest smoke in Fulpmess, Austria.

The effect on springtime loss of the snowpack in forests can vary from one to three weeks. And because the melting and evaporation process is prolonged, there is more opportunity for the soil to absorb the water. Runoff is reduced and with it the chance of flooding.

Even when there is runoff, forests are instrumental in preventing and reducing soil erosion. Forested regions inhibit soil erosion for slopes up to 20° to 30° steep. In comparison, roads begin to erode when the slope is in the range of 5° to 10°, while open fields erode because of surface runoff when the slope is quite gentle—1° to 7°.

Weather Effects of Clearings, Edges, and Windbreaks

Clearings Depending on a clearing's size, weather conditions in a clearing can vary from those in the forest to those over open terrain. Although virtually all weather elements will undergo some amount of modification or change from forest to clearing, two elements can change enough to affect the choice of a campsite and the comfort of the camper. The two are temperature and wind.

Nighttime temperatures in the forest remain warm relative to air over an extended, open area. The chances for cold temperatures and frost increase as the clearing becomes larger. The effect of the clearing on temperature can be judged by considering the diameter (D) of the cleared area and the height (H) of the surrounding trees. As the ratio D:H gets larger, minimum temperatures in the center of the clearing become increasingly like those over open terrain. Where the clearing is twice as wide as the trees are tall (D:H = 2), there is little difference between the center of the clearing and open terrain: When D:H = 1.25 or less, clearing temperatures resemble those inside the forest. Daytime temperatures are also affected by the ratio of D:H. There is an increase in the penetration of solar energy as the clearing is enlarged, while wind flow into the opening also increases with larger value of

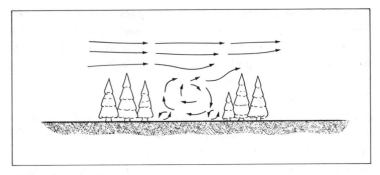

Airflow in a clearing.

D:H. The resulting effect on peak daytime temperatures is more complex than at night: Daytime temperature maximums increase until D:H reaches a value of 1.8; thereafter, peak temperatures decline with further increases in the width of the clearing.

Airflow in a clearing is important to the camper. When the wind is moderate or stronger above the treetops, an eddy will form in the clearing similar to those that form in the lee of hilly and mountainous terrain. The downwind portion of the clearing will have intermittently gusty winds with a strong downward component. Campfires built in this area are apt to create smoky conditions at the campsite as the downdrafts force the smoke down and to the center of the clearing. On the other hand, the draft is improved for fires built in the windward sector where updrafts prevail. If the forest is dense enough at the edges of the clearing, secondary eddies will form along both the upwind and downwind edges. These areas should be avoided when making a fire, although they do offer moderate nighttime temperatures.

Edge Effects Several weather effects found at the edge of the forest have already been discussed or implied: the formation of wind eddies both on the windward and leeward edges; temperature (and humidity) variations; and the filtering of fog droplets.

The braking effect the forest has on the wind takes place almost totally along the edge of the forest. The eddy upwind of the forest edge induces the air to flow up and over the trees rather than into the woods. What wind does penetrate the forest is quickly slowed down by the undergrowth and thick array of tree trunks. When the wind attacks the forest edge at angles larger than 45°, the wind is turned by the forest "wall" in such a way that it blows parallel to the boundary. The wind speed increases markedly in the open area adjacent to the edge of the forest; hiking or camping in this area may be quite unpleasant.

Perhaps the most beneficial effect of the forest is its cleansing of the air of most of the suspended dust particles. The particles that float in the air for long periods of time are very small; they also pose the largest potential danger to health because their small size allows them to pass through the body's natural filtering membranes into the respiratory tract and lungs. The forest is a very effective outdoor filter. The low wind speeds enable many of the particles to settle quickly out of the air, while others are deposited on needles, leaves, and other plant surfaces. The forest is so effective that about 85 percent of the dust particles in an approaching airflow are removed after traveling only 100 yards through the forest.

Shelterbelts and Windbreaks Regions where the wind persistently blows from the same direction are often protected by barriers called *shelterbelts* or *windbreaks.* These barriers typically consist of long, narrow, tall stands of trees planted perpendicular to the prevailing winds. Another type of windbreak is the snow fence, used to prevent blowing snow or sand from drifting across or piling up on a roadway or other sensitive area. Shelterbelts are effective barriers to the *prevailing* winds; they are not intended to provide protection against unusually strong winds, such as gales, hurricanes, or tornadoes. Two things happen in the vicinity of a shelterbelt that cause the wind near ground level to be reduced for long distances behind the barrier. First, wind eddies are avoided in the lee because the porous shelterbelt is not a solid obstruction to the wind flow,

Agricultural shelterbelts or windbreaks cause significant reductions in the wind speed, thereby minimizing soil erosion and wind damage to new crops. Courtesy of Walter F. Dabberdt

allowing some air to pass through. Second, the shelterbelt deflects the major portion of the airstream up and over the barrier, where the air becomes accelerated. Since there is no eddy circulation on the downwind side of the shelterbelt, the accelerated stream of air remains aloft until it loses its identity by gradually mixing into and with the unaffected air farther downstream.

Shelterbelts are remarkably effective. Thin and moderately thick windbreaks are significantly better than very thick ones. The ideal shelterbelt allows 40 to 50 percent of the air to penetrate through. Multiple rows of tall, thin trees (at least three) having a total width of 5 to 10 yards seem to work particularly well when large areas need protection. Near the ground, the lowest wind speeds usually are *not* found immediately behind the windbreak. Instead, minimum wind speeds occur up to five shelter-heights downwind. The exception to the rule is the very dense shelterbelt; it reduces the wind speed just behind the barrier to about 15 percent of its value in the open; however, the wind speed recovers rapidly and is back to 60 percent at five shelter-heights and 80 percent at ten shelter-heights.* The moderately dense shelterbelt does a better job of diminishing the wind speed over a longer distance: The minimum wind speed is 35 percent of the upwind value and occurs about four shelter-heights downwind; at ten shelter-heights downwind, the speed is still only 60 percent of the upwind value in the open.

The area influenced by a shelterbelt seems to depend only on the height and thickness of the barrier. For all practical purposes, a 50-foot-tall windbreak will reduce the wind speed 250 feet downwind in the same way that a 5-foot-tall hedge influences the wind 25 feet downwind. When a snow fence is used for snow or sand control, the maximum amount of snow or sand will accumulate in the region where the wind speeds are lowest—four or so shelter-heights downwind. Lesser amounts will accumulate both farther upwind and downwind, corresponding to the reduction in the wind speed. As the snow or sand dune builds, its own height and shape eventually exert more influence on the wind than the fence that created it.

A Norwegian meteorologist once told me of an interesting experience with snow fences and how not to use them. It was the early 1940s and the occupation troops in Norway were intent on keeping a certain highway open during the coming winter. A group of nationals was assembled to do the manual labor under direction of the troops. The locals started erecting snow fences about 8 yards to either side of the roadway. The troops, suspecting some subversive intent, forcefully insisted that the fences be located in

---

*The wind speeds cited refer to a height nearly 5 feet above ground level; closer to the ground the wind reductions are larger and extend farther downwind.

a more effective place—immediately adjacent to the highway. The locals complied, the fence was erected, and by the end of the first snowstorm the road was closed for the winter.

Shelterbelts have been used successively in the midwestern and plains states for control of soil erosion. When multiple barriers are used, they are separated horizontally at distances approximately 25 times the height of the individual barriers. If the trees that are used mature at 50 feet, the windbreaks are placed no more than one-quarter mile apart. Although 2 to 2.5 percent of the available land is used for the shelterbelts, the benefits from reduced soil erosion (and evaporation) outweigh the costs.

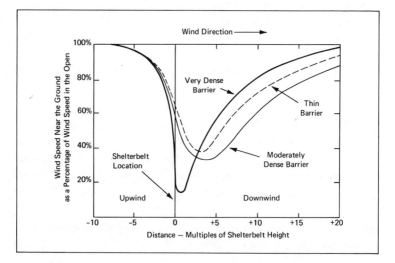

The density of a shelterbelt determines the magnitude of the reduction in wind speed; the height of the shelter determines the extent of the reduction.

Forest Fires

Weather plays a significant role in the creation of forest fires, both directly and indirectly. Direct effects include the roles of wind, temperature, lightning, and stability in the ignition, spread, and intensity of wildland fires. An equally important, if somewhat indirect, effect is the influence of weather elements on fuel moisture and consequently the flammability of dead and living fuels.

Direct Weather Effects   Thunderstorms are a major cause of forest fires, particularly when the storms are "dry." This happens when the precipitation falling out of the cumulonimbus cloud evaporates before reaching the ground. With the ground dry, lightning easily starts numerous fires (especially in the late summer or toward the end of extended dry periods). Thunderstorms are doubly dangerous because the downdrafts and gusty winds transport the heat and burning embers and replenish the necessary supply of oxygen. The western forests are particularly dry and especially susceptible to lightning fires toward summer's end, when steady rains may not have fallen for four or five months. It is not uncommon for several hundred fires to start during a single day in one forest, burning up to hundreds of thousands of acres in a few short days.*

All of the winds unique to mountains, valleys, and shorelines are important to the spread of a fire. They also help control the evaporation process and with it, the flammability of the forests. After a fire has started, the role of even local wind circulations cannot be underestimated. As an example, there are reports of forest fires progressing downhill with nocturnal downslope winds in spite of the enormous amounts of heat released. This type of spread is contrary to intuition, which would suggest upslope movement because of the convection currents associated with the fire.

Other weather elements are equally important: Temperature

---

*100,000 acres is equivalent to about 156 square miles, or a square area about 12½ miles on a side.

A cumulus congestus cloud caps an intense forest fire. Courtesy of the U.S. Forest Service

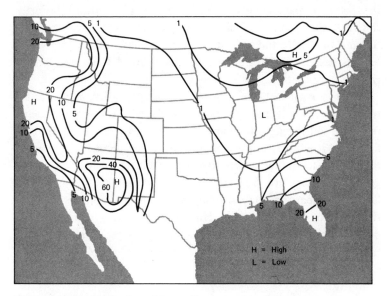

Yearly average distribution of forest fires caused by lightning

directly affects the flammability of the fuel; warmer temperatures reduce the amount of heat required to start and sustain ignition. Clear skies and low humidities stimulate evaporation, further diminishing the heat necessary to ignite or intensify a fire. Atmospheric stability also plays a controlling role. Stable conditions (inversions, for example) suppress vertical air movement and the rate at which oxygen is supplied to the fire. On the other hand, unstable conditions favor strong up- and downdrafts and ensure a plentiful oxygen supply.

Indirect Weather Effects—Fuel Moisture   When referring to fuel, we include all organic material—dead and alive (or green). *Fuel moisture* is the ratio of the water content of the fuel to the *dry* weight of the fuel. Dead fuels typically have moisture levels in the range of 1.5 to 30 percent, while green fuels are usually moister (3 to 200 percent). Fuel moisture is an important determinant of the potential flammability of any fuel. Simply stated: The drier the fuel, the lower its fuel moisture content and the higher its flammability potential. Moist fuels have lower flammability potentials because it takes additional quantities of heat to evaporate the moisture in them.

Water is essential to living plants and explains their higher moisture levels. Water is used internally by plants to transport nutrients from the soil through the plant; it is used in the leaves to produce food; and it is used to transport manufactured products to growing tissues and to storage points. Because of high fuel moisture levels, ground fires are not a serious threat to deciduous forests in full leaf. By comparison, the live foliage of conifers is more combustible than that of deciduous trees. However, any living vegetation can be consumed by fire. The intensity of the fire is diminished as the fuel moisture content becomes larger. A sizable fraction of

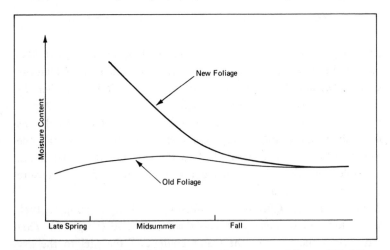

Seasonal variation of the fuel moisture of evergreen foliage. *Old foliage varies but slightly throughout the year, whereas new foliage has very high moisture content in the late spring–early autumn that tapers off to match that of the old foliage by midfall.*

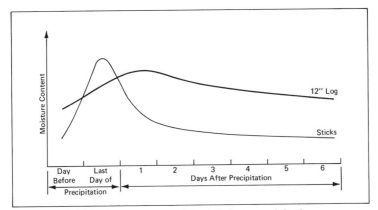

Response of the moisture content of dead fuels to precipitation

the heat given off in combustion must then be used to vaporize the water, thereby decreasing the heat available to ignite new fuels.

Dead trees and other dead fuels retain the ability to take up and give off water. They may do this by absorbing liquid water, and can acquire fuel moisture levels two to three times their dry weight. However, the rate of absorption is slow and the standing water may evaporate before the log becomes saturated. Dead fuels may also take up water directly from a moist atmosphere by the process of adsorption. The maximum amount of moisture obtained this way falls in the range of 30 to 35 percent of the dry weight of the fuel. When the vapor pressure of the air becomes less than the vapor pressure of the water bound in the dead wood, the fuel gives up moisture to the air.

The flammability of a particular forest depends on a wide range of factors: the types and abundance of dead and green fuels, past weather and climatic conditions, and current local and regional weather patterns. In this way, the effects of a previous year's drought may be equally or more important than today's weather. Differences in the flammability potential will also vary with slope and elevation in response to local patterns of wind, temperature, humidity, sunshine, cloudiness, and precipitation.

Postscript

The weather picture painted in Part Two is both simple and complex. There are likely to have been a number of examples of weather in the mountains and forests and on the water that completely describe the elements as you encountered them yesterday or will find them tomorrow. But there may be an equal number (or more) of weather conditions that don't fit the mold we have formed. It may be that the large-scale weather pattern totally dominates the local effects of the trees, terrain, or open waters. It's equally possible, however, that the weather effects are indeed purely local in their origin, but that you don't quite recognize them. Hiking on a mountain slope blanketed by a tall stand of ponderosa pine or sugar maple, you have to integrate everything you know about mountain effects with your knowledge of weather changes brought about by forests. Once you start thinking about places you have been, you'll be surprised by the number that have their weather influenced by a combination of terrain slope, tree cover, and large expanses of water. Although the weather effects of each have been discussed separately in Chapters 6 through 8, they are at work at the same time in the same mass of air to produce the weather conditions you experience. Sometimes they work together, but often one will oppose the other in an atmospheric tug of war.

By no means should you think of the information presented in Parts One and Two as the end of your personal weather primer. Far from it. This is only the beginning. The true test will come in the wilds when you match wits with the weather machine. You'll be tested over and over again. And when you think you've got the upper hand at anticipating nature's vagaries, you can rest assured that what you really know is that you don't know—at least not always.

The important first step toward understanding and forecasting the weather and how it will affect you is to be attuned to the air around you. Benjamin Franklin realized more than two centuries ago that most people are oblivious to weather, its causes, and its changes when he penned the now well-known rhyme:

> Some people are weatherwise,
> but most are otherwise.

Join the minority!

Appendix A Glossary

ABSOLUTE TEMPERATURE   A temperature scale with its zero point equal to the temperature at which all molecular motion ceases. Water freezes at 273.16 degrees Absolute and melts at 373.16 degrees Absolute. Temperature *changes* on the absolute scale are equal to those on the Celsius scale.

ADVECTION   Horizontal transport (for example, of heat or moisture) by the wind.

AIR MASS   A widespread body of air that is at least several hundred miles across and that has relatively uniform temperature and humidity characteristics.

ALTIMETER   An instrument used to measure altitude or elevation. The two basic types are: the *pressure altimeter,* which uses an aneroid barometer to measure pressure from which altitude is estimated; and the *radio altimeter,* which works on the radar principle.

ANEMOMETER   An instrument that measures the speed of the wind.

ANTICYCLONE   A high-pressure weather system (high) characterized by a clockwise flow of air and usually fair weather.

ATMOSPHERE   The envelope of air that surrounds the earth.

AURORA   Light emitted from the upper atmosphere over the middle and upper latitudes. Caused by the bombardment of the upper atmosphere by charged particles emitted from the sun, the colored displays variously resemble blue, green, or reddish curtains, bands, rays, and arcs.

BACKING   A counterclockwise shift of the wind; for example, from east to north.

BAROMETER   An instrument for measuring atmospheric pressure. The two principal types are the *aneroid barometer,* which has a partially evacuated capsule that flexes as pressure changes; and the *mercurial barometer,* which has an evacuated tube of mercury whose height increases with increasing atmospheric pressure.

BEAUFORT SCALE   A system invented in the early nineteenth century by British admiral Sir Francis Beaufort to estimate wind at sea and on land.

BLIZZARD   A severe winter storm with low temperatures, heavy snows, and high winds.

BLOCKING   A large-scale weather situation in which the normal west-to-east movement of highs and lows is stalled or slowed down, causing weather conditions to remain stationary across large portions of the country (although winds often persist within the pressure systems).

CALM   The absence of a detectable wind (generally less than 1 mph).

CEILING   The height above ground of the bottom of the lowest layer of a cloud that obscures more than half of the sky.

CELSIUS   A temperature system widely used in scientific work and in most countries outside the United States in which the freezing point of water is 0°C and the boiling point is 100°C.

CHINOOK   A local name for the foehn; a strong, warm, and dry wind on the eastern side of the Rocky Mountains.

CLIMATE   The long-term average of weather elements; usually a period of 30 years is used to determine average conditions.

CLOUD   A collection of small water droplets or ice particles that form a visible mass. A cloud has its base above ground level, whereas fog is a cloud that touches the earth's surface.

COLD FRONT   The boundary between a passing warm air mass and an advancing colder air mass. Marked changes occur in temperature, humidity, wind, and precipitation with passage of the front.

CONDENSATION   The process by which gaseous water vapor turns into liquid water droplets because of the addition of water vapor or the lowering of the air temperature, or a combination of the two.

CONVECTION   Updrafts created when air temperatures decrease sharply with altitude.

CORIOLIS EFFECT   An apparent force caused by the earth's rotation that causes winds in the Northern Hemisphere to veer to the right as they travel, although the speed of the wind is unaffected. The force acts as well on flying missiles, railroad trains, and other moving objects.

CYCLONE   Another term for a low-pressure storm system, although it is also popularly (but improperly) used to describe tornadoes, waterspouts, and dust devils.

DEEPEN   To decrease atmospheric pressure, as in an intensifying low-pressure system.

DEPRESSION   A region or area with low atmospheric pressure at its center.

DEW   Water droplets that have formed by condensation on grass and other objects on or near the ground.

DEW POINT   The temperature to which air of a given pressure and moisture content must be chilled for dew to form.

DOLDRUMS   An oceanic region near the equator where the winds are usually light and variable.

DRIZZLE   Small water droplets that appear to float in the air, but which actually fall gently to the ground to produce precipitation at a rate less than about one-twentieth of an inch per hour.

DRY BULB Refers to either the dry-bulb temperature or a dry-bulb thermometer, which are identical with air temperature and the common thermometer, respectively. The term is usually used to denote the unwetted thermometer found in a sling psychrometer.

EVAPORATION The process by which water is converted from a liquid to a gas (water vapor).

FAHRENHEIT The temperature scale commonly used in the United States in which the freezing point of water is 32°F and the boiling point is 212°F.

FILL To increase atmospheric pressure, as in the weakening of a low-pressure system.

FOEHN A warm, dry, and often strong wind frequently found on the downwind side of mountain ranges; more commonly called a chinook in the Rocky Mountains.

FOG A cloud that touches the ground and reduces the visibility to less than 1 km (.6 mile).

FORECAST A prediction of weather conditions less than three days ahead; predictions for longer lead times are less reliable and are called outlooks.

FREEZING LEVEL The altitude at which the air temperature first drops to 32°F.

FRONT The boundary zone between two dissimilar air masses; marked by abrupt changes in wind, temperature, humidity, pressure, and precipitation from one side of the frontal zone to the other. Fronts may be very narrow and abrupt, or fairly wide (tens of miles) and transitional.

FROST A deposit of ice crystals on unsheltered objects caused by sublimation of water vapor.

FROST POINT The temperature to which air of a given pressure and moisture content must be chilled for frost to form; the frost point must be less than 32°F.

FUNNEL CLOUD The name of a tornado cloud that does not

touch the ground; the visible funnel cloud is made up primarily of water droplets, whereas a tornado cloud also consists of dirt and debris.

GALE   An unusually strong wind. Specifically, a wind of at least 28 knots and not more than 55 knots.

GLAZE   A thin, clear coating of ice caused by the rapid freezing of a thin film of water on cold objects.

GUST   A sudden, brief increase in the speed of the wind.

HAIL   Irregularly shaped lumps of ice formed within convective (usually cumulonimbus) clouds.

HALO   Colored or whitish rings around the sun or moon created when their light passes through a thin ice-crystal cloud.

HAZE   Small dust, salt, or pollution particles that are dispersed through a portion of the atmosphere, causing a noticeable reduction in visibility.

HIGH   A common term for an anticyclone, or a region with relatively high atmosphere pressure at its center.

HUMIDITY   The amount of moisture present as a gas (water vapor) in the air.

HURRICANE   An intense low-pressure storm system that forms in the tropical oceans, characterized by heavy rains and strong winds (greater than 64 knots).

INVERSION   A condition in which air temperatures become warmer with increasing height above ground.

ISOBAR   A line on a weather map joining points (places) with the same value of atmospheric pressure.

ISOTACH   A line on an aviation weather map joining points with the same value of wind speed.

ISOTHERM   A line on a weather map joining points with the same value of air temperature.

JET STREAM   A meandering narrow stream of very strong winds found at altitudes ranging from 30,000 to 50,000 feet.

KNOT   A unit of speed equal to one nautical mile per hour or about 1.15 mph.

LAND BREEZE   Nocturnal wind that blows offshore because of the difference in temperatures between land and water.

LIGHTNING   A visible electrical discharge produced by a thunderstorm.

LOW   A common term for a cyclone, or a region with relatively low atmospheric pressure at its center; usually characterized by weather fronts and poor weather.

METEOROLOGY   The study of the atmosphere, including the understanding, prediction, and control of weather.

MILLIBAR   A unit of atmospheric pressure. The average sea-level pressure is equal to 1,013.2 millibars.

MIRAGE   An optical effect caused by large variations in the density of the air with height; can cause objects to appear distorted, displaced, or inverted.

MIST   Similar to drizzle, mist consists of very small water droplets that produce a thin, grayish veil that is, however, less dense than fog.

MONSOON   Seasonal winds that blow between large land masses and adjoining oceans because of the difference in temperature and pressure between the two.

OCCLUDED FRONT   A weather front that is formed when an advancing cold front overtakes either a warm front or a stationary front.

PRECIPITATION   Water or ice particles that fall to the ground, such as rain, drizzle, hail, and snow.

PRESSURE (atmospheric)   The force exerted on a unit area (for example, a square inch) of an object from the weight of the overlying atmosphere.

PRESSURE GRADIENT   The change in pressure between two points some distance apart.

PRESSURE TENDENCY   The change in pressure with time at a single location.

PROBABILITY   The chance or likelihood that something will happen. "A 10 percent probability of rain tomorrow" means that there are only 10 chances in 100 that it will rain, or that the same combination or pattern of weather elements is likely to produce rain only 10 times in 100.

PSYCHROMETER   A device consisting of a dry-bulb and a wet-bulb thermometer, used to determine dew point and relative humidity, provided atmospheric pressure is also known or estimated.

RADIOSONDE   An instrument that measures temperature, humidity, and pressure; it is usually carried through the atmosphere by a weather balloon and transmits its observations back to ground.

RELATIVE HUMIDITY   The amount of moisture present in the air, expressed as a percentage of the moisture the air could hold at its current temperature.

RIDGE   A tongue-shaped region of relatively high atmospheric pressure as outlined by the isobars on a weather map. Unlike an anticyclone, the isobars around a ridge do not form a closed circular pattern.

RIME ICE   A milky, granular deposit of ice that forms from the rapid freezing of very cold raindrops as they fall onto a subfreezing object.

SCUD   Low-level, small and raggedy cloud elements that appear to have been torn away from either stratus or cumulus clouds.

SEA BREEZE   An onshore, daytime wind that blows near the coastline and is due to the temperature differential between a cool water surface and a warm land surface.

SHOWER   Sudden and often intense precipitation from cumulus clouds.

SLEET   Ice particles formed by the freezing of raindrops when falling through a subfreezing atmospheric layer.

SMOG   A term popularly used to denote visible air pollution produced by chemical reactions in the air among industrial, residential, vehicular, and often natural sources of emissions.

SQUALL LINE   A long, narrow band of intense thunderstorms not associated with a weather front.

STATIONARY FRONT   The boundary between two immobile air masses. When one air mass displaces the other, the stationary front is then called either a warm front or a cold front, depending on the relative warmth of the trailing air mass.

STRATOSPHERE   A region of the atmosphere above the troposphere and below the mesosphere, typically extending from 45,000 feet to about 80,000 feet.

SUBLIMATION   A process by which water is transformed directly from the ice (solid) phase to the vapor (gas) phase, or vice versa, without going through the water (liquid) phase.

SYNOPTIC CHART   A weather map.

TORNADO   A violently rotating column of air (often called a funnel cloud) that descends from a severe cumulonimbus (that is, thunderstorm) cloud. The funnel typically has a diameter of several hundred yards or more with very low pressures at the center and wind speeds that range from 100 to 300 mph.

TROPOSPHERE   The lowest layer of the atmosphere, extending from the earth's surface to the tropopause (the boundary between the top of the troposphere and the base of the stratosphere). The troposphere typically varies in thickness from 8 to 10 miles.

TROUGH   A tongue-shaped region of relatively low atmospheric pressure as outlined by the isobars on a weather map. Unlike a cyclone, the isobars do not form a closed circular pattern.

TWISTER   Common term for a tornado.

TYPHOON   Name given to hurricanes in the western Pacific Ocean.

VEERING   A clockwise shift in the direction of the wind, for example, from north to east.

VIRGA   Precipitation that can be seen falling from a cloud but that evaporates before reaching the ground.

VISIBILITY   The farthest distance at which a dark-colored object on the horizon can be seen against the sky.

WARM FRONT   The boundary between a passing cool or cold air mass and an advancing warmer air mass. With passage of the front, the wind veers and temperatures rise. Steady rain usually occurs ahead of and during passage of the front.

WARM SECTOR   A wedge-shaped mass of warm air bordered on the east by a warm front and on the west by a cold front.

WATERSPOUT   A tornadolike cloud over water. Waterspouts are less intense and less damaging than their land counterparts.

WEATHER   Meteorological conditions as they exist at any point in time (distinct from climate, which is the long-term average of weather). Frequently, the term "weather" is used to denote only poor or stormy conditions.

WET BULB   Refers to either the wet-bulb temperature or a wet-bulb thermometer. The wet-bulb thermometer consists of a common thermometer with a wetted wick wrapped around its mercury (or alcohol) bulb; the wet-bulb temperature is obtained by twirling the thermometer, causing the water to evaporate and the temperature to fall below the air temperature. Together, the wet-bulb and air (dry-bulb) temperatures are used to measure the dew point and relative humidity.

WIND   Air in motion.

WINDCHILL   The cooling effect of the wind that results from the loss of body heat due to evaporation of moisture from the surface of exposed skin. The combined effects of cold air and wind are often expressed by an equivalent air temperature—the windchill temperature.

WIND DIRECTION   The direction from which the wind blows, reported either as points of the compass (for example, easterly) or degrees (for example, 090°).

Appendix B Weather and Your Health and Comfort

Weather conditions can affect your health and comfort by causing body temperatures to rise above or fall below the normal temperature of 98.6°F. The three weather elements that have the greatest effect are temperature, wind speed, and relative humidity. Individually and in combination they control the rate at which the body must produce energy to remain warm in cold conditions and the ability of the body to dissipate excess heat to remain cool in a warm environment.

There are five ways that the body exchanges heat with its environment: evaporation, conduction, convection, radiation, and respiration. Evaporation depends on the relative humidity of the air, wind speed, and the moistness of the skin. Heat is conducted between the body and the ground surface wherever the two are in contact and whenever they are at different temperatures. Convection transfers heat from the warm skin surface to the cooler air, particularly when there is some wind. As with all warm bodies, heat is lost by giving off infrared radiation and is received by absorbing solar energy. The top of a bare head can be particularly vulnerable to significant heat loss in cold conditions. Heat is also lost by the warm air we exhale. Wrapping a scarf around the mouth during a cold spell transfers some of this heat to the wool fibers and the air entrapped between scarf and face. This in turn helps to prewarm the air inhaled, thereby reducing the heat the body must produce to keep itself warm.

Hypothermia is the loss of body heat, resulting in the lowering of body temperature. In the early stages, blood flow to the skin surface is automatically decreased by the body in an attempt to minimize heat loss and the subsequent lowering of the temperature of the body core. At the same time, however, the danger of frostbite increases. With decreasing body temperature, there are increasingly severe physical impairments and hazards, ranging from impairment of the ability to think clearly to helplessness, unconsciousness, and ultimately even death.

In cold conditions the effects of temperature and wind are inseparable; they act together to cause a loss of body heat. Humidity is not particularly important, except for the effect it can have on the insulation properties of certain garments and for the critical effects due to wet skin or clothing.

The combined effect of low temperatures and increasing winds can be gauged by the *windchill factor* (see Table 26, page 225). It is an equivalent temperature that describes the freezing effects of cold, moving air on exposed flesh. The windchill factor expresses this effect in terms of the *calm air* temperature that has the same freezing effect. Bear in mind that the air can be still, but you may be moving. This is the case for skiers and snowmobilers. In these cases, your traveling speed replaces the actual wind in determining the windchill factor. To be totally precise, the apparent wind speed should be used as was discussed in Chapter 6, "Lakes and Oceans."

Even more critical than high winds is the effect of water chill due to wet clothing and moist skin. Body heat is lost rapidly by conduction from skin to clothing and then by convection and evaporation. The insulation value of most wet clothing decreases by 90 percent. Moist skin loses heat by evaporation, compounding the dangers. For these reasons there is a very simple and effective rule for survival in cold environments: Do not sweat!

HYPOTHERMIA (LOSS OF BODY HEAT) SAFETY RULES

1. Know and respect the conditions that can cause hypothermia—cold, wind, and wetness.
2. Interpret the weather forecast to evaluate the danger of hypothermia.
3. Prepare yourself for adverse weather by wearing or carrying warm, dry clothing and water-repellent garb; carry a tube tent, poncho, large leaf bag, or the like that can be used as an emergency shelter.
4. Make camp early if you know a storm is coming; exposure to the elements can quickly sap your strength.
5. Keep your muscles moving so that your body can produce heat. Do not work up a sweat, as this will reduce the insulation value of your clothing and enhance heat loss by evaporation and conduction. Refuel your body by eating high-energy, sugary foods.

TABLE 25. EFFECTS OF HYPOTHERMIA

Body Temperature (°F)	Symptoms and Effects
98–96°	Shivering becomes intense and uncontrollable; hands become numb, and complex tasks are performed with some difficulty.
95–91°	Shivering becomes violent and speaking is difficult; ability to help yourself is impaired.
90–86°	Muscles become rigid and shivering decreases. Too late for self-help as thinking is impaired. Victim may, however, still appear erect and in touch with surroundings.
85–81°	Muscular rigidity progresses, and victim becomes increasingly irrational and irresponsive. Pulse and respiration rate decrease.
80–78°	Victim becomes unconscious and most reflex actions cease. Heartbeat is erratic.
less than 78°	Death is virtually inevitable.

Hyperthermia is the opposite of hypothermia; it is an excess of body heat that can be caused or exacerbated by high temperatures, low wind, and high humidity. Under these conditions there normally is abundant sunlight, which supplies the body with unwanted heat. High humidity and low wind shut off the body's evaporative cooling system. If air and ground are above 99°F, the body will gain heat by conduction and convection. With this gloomy and inhospitable environment, the body has only limited abilities to lose excess heat. Even that lost by respiration is more than offset by the hot air that is inhaled. Some heat is lost by radiation, but the burden falls primarily on evaporation. But losing too much water can result in dehydration and electrolytic imbalance. It is crucial that you drink water as you become thirsty, consuming a small amount of salt at the same time. Bodily heat production should be reduced by minimizing muscular movements and avoiding or eliminating consumption of high-protein foods.

There are four stages or types of heat problems you may encounter in the outdoors: *Heat weakness* is the first and mildest stage. It is caused by a hot and humid environment, and results in fatigue, headaches, poor appetite, insomnia, heavy sweating, a rapid pulse, and loss of strength. *Heat cramps* result from strenuous activity in the same environment that produces heat weakness. The very heavy sweating depletes salt levels in the body, causing intermittent leg or abdominal cramps. *Heat exhaustion* is more serious and results from prolonged exertion during hot and humid weather. The body's heat regulatory system fails, causing a problem with the circulatory system as well. The victim may vomit or become delirious. Blood pressure usually is low and pulse erratic. The skin feels moist and clammy. In the case of *heat stroke,* automatic cooling from sweating stops and the body temperature can quickly jump to 106° and higher. The victim may lapse into a coma, while the pulse rate is rapid and blood pressure is high.

The key to avoiding these heat problems is to avoid the weather conditions and physical activities that induce them.

TABLE 26. WINDCHILL FACTOR

Actual Temperature (°F)	Wind Speed									Level of Danger That Exposed Flesh Will Freeze
	0–4	5	10	15	20	25	30	35	40	
45	45	43	34	29	26	23	21	20	19	
40	40	37	26	23	19	16	13	12	11	
35	35	32	22	16	12	8	6	4	3	Little danger
30	30	27	16	9	4	1	−2	−4	−5	
25	25	22	10	2	−3	−7	−10	−12	−13	
20	20	16	3	−5	−10	−15	−18	−20	−21	
15	15	11	−3	−11	−17	−22	−25	−27	−29	
10	10	6	−9	−18	−24	−29	−33	−35	−37	
5	5	0	−15	−25	−31	−36	−41	−43	−45	Increased danger
0	0	−5	−22	−31	−39	−44	−49	−52	−53	
−5	−5	−10	−27	−38	−46	−51	−56	−58	−60	
−10	−10	−15	−34	−45	−53	−59	−64	−67	−69	
−15	−15	−21	−40	−51	−60	−66	−71	−74	−76	
−20	−20	−26	−46	−58	−67	−74	−79	−82	−84	Severe danger
−25	−25	−31	−52	−65	−74	−81	−86	−89	−92	
−30	−30	−36	−58	−72	−81	−88	−93	−97	−100	

The windchill factor is the temperature the body feels as a result of the combination of the actual air temperature and the chilling effect of the wind. To use the table: First, find the row with the actual air temperature. Next, choose the column heading that corresponds to the wind speed. The equivalent windchill temperature lies at the intersection of the column and row. For example, air at 0°F with a 15-mph wind has the cooling effect of −31°F under calm conditions. The effect is identical if the air is calm but you are moving.

TABLE 27. TEMPERATURE-HUMIDITY DISCOMFORT

Actual Temperature (°F)	Relative Humidity (%)									
	10	20	30	40	50	60	70	80	90	100
70	C	C	C	C	C	C	C	C	C	C
72	C	C	C	C	C	C	C	MD	MD	MD
74	C	C	C	C	C	MD	MD	MD	MD	MD
76	C	C	C	C	MD	MD	MD	MD	MD	MD
78	C	C	C	MD	MD	MD	MD	D	D	D
80	C	C	MD	MD	MD	MD	D	D	D	WD
82	C	MD	MD	MD	MD	D	D	D	WD	WD
84	MD	MD	MD	MD	D	D	D	WD	WD	WD
86	MD	MD	MD	D	D	D	WD	WD	I	I
88	MD	MD	D	D	D	WD	WD	I	I	I
90	MD	D	D	D	WD	WD	I	I	I	I
92	D	D	D	WD	WD	I	I	I	I	I
94	D	D	WD	WD	I	I	I	I	DZ	DZ
96	D	D	WD	WD	I	I	I	DZ	DZ	DZ
98	D	WD	WD	I	I	I	I	DZ	DZ	DZ
100	D	WD	I	I	I	I	DZ	DZ	DZ	DZ
102	WD	WD	I	I	I	DZ	DZ	DZ	DZ	DZ
104	WD	WD	I	I	I	DZ	DZ	DZ	DZ	DZ
106	WD	I	I	I	DZ	DZ	DZ	DZ	DZ	DZ
108	WD	I	I	I	DZ	DZ	DZ	DZ	DZ	DZ
110	I	I	I	DZ	DZ	DZ	DZ	DZ	DZ	DZ

Key:

C—Comfort zone; relatively few people are uncomfortable

MD—Moderate discomfort; up to one-half the population is uncomfortable

D—Discomfort; nearly everyone is uncomfortable

WD—Widespread discomfort; acute discomfort affects work

I—Impairment; decrease in efficiency and increase in accidents and mistakes

DZ—Danger zone; potentially critical impacts on everyone

SAFETY RULES FOR HOT ENVIRONMENTS

1. Estimate how long the emergency situation is likely to last and plan accordingly, allowing some margin of safety.
2. Stay out of the direct sunlight. Avoid hot ground surfaces and light-colored ones that reflect large percentages of sunlight. Construct an improvised shade or shelter if no natural ones exist.
3. Slow down bodily heat production by minimizing muscle movement. Avoid foods high in protein. In fact, avoid eating unless the emergency persists for a long time.
4. Conserve body water. Minimize or avoid sweating. Keep covered with light-colored, lightweight clothing. Drink water when thirsty and consume small amounts of salt.
5. Work and travel at night.

TABLE 28. SURVIVAL AND WATER SUPPLY IN A HOT ENVIRONMENT

Water Supply	Survival Time While Only Sitting	Distance You Can Travel at Night (sitting during daytime)	Survival Time, Traveling Nights Only
None	3 days	10–15 miles	2 days
1 gallon	4 days	20–30 miles	3 days
2½ gallons	5 days	30+ miles	3½ days

Appendix C

Instant Weather Information

Up-to-date weather information is now available to nearly all outdoorsmen. Whether you are ocean fishing or mountaineering, houseboating or hunting, weather forecasts and weather observations are at your disposal through two special weather radio systems: the Transcribed Weather Broadcast Service (TWEB) provided by the Federal Aviation Administration (FAA), and the radio weather broadcasts of the National Oceanic and Atmospheric Administration (NOAA). Together, these two radio networks provide weather information to most, but not all, portions of the country. The information each provides is comprehensive, current, and available at all hours of the day. However, the radio frequencies are not within the AM/FM bands covered by regular commercial radio, and so you will need to be equipped with a multichannel radio or a special radio designed specifically to receive the NOAA broadcasts.

The NOAA weather radio reports are broadcast from over 300 locations that are within range of 90 percent of the U.S. population. Three frequencies are used throughout the network: 162.40 megaherz (MHz), 162.55 MHz, and 162.475 MHz. These VHF broadcasts are above the commercial FM band, which ends at 108 MHz. Radio weather transmissions have a fairly limited range, however, and usually have an effective broadcast radius of 40 miles or less. The actual range at any location depends on transmitter characteristics, local geography, and the sensitivity of your radio.

NOAA VHF weather broadcasts provide a wealth of weather information, in plain and understandable language, ranging from a full summary of local weather conditions at observing stations throughout the area to locally tailored short- and long-range weather forecasts. The four- to six-minute broadcasts are repeated continuously, twenty-four hours a day. They are revised every two to three hours, or more often if required. Special information is frequently provided depending on the needs of the local area, such as advisories to motorists, campers, and sportsmen. Along the coasts, for example, this special information includes sea and swell conditions, small craft warnings, tidal conditions, visibility, and so forth.

Apart from issuing routine weather information, another major purpose of the radio weather broadcasts is to provide emergency weather information. Thus, when severe weather threatens, routine broadcasts are interrupted and emergency warnings are given. At the same time that an emergency warning is given, a special signal is broadcast that can automatically trigger an alarm mechanism on the receiving radio. Not all radios are equipped with the alarm feature. Several types of alarms are available. Some consist of a flashing light or siren; others cause the radio to automatically turn itself on at full volume. The cost of these special radios is quite reasonable, especially when the value of the information provided is considered.

If you are not within the range of NOAA weather radio, there is a good chance you may be within range of the so-called air weather broadcasts that are prepared by NOAA and transmitted by the FAA. There are over 100 TWEB stations on the mainland, and each has an effective broadcast range of 75 miles. The broadcast sites are usually located at the larger airports. Most of the transmissions are made using the longwave band with frequencies in the range of 200 to 415 KHz (although ten of the stations broadcast in the 108–118 MHz band).

The primary purpose of the air weather broadcasts is to provide weather information for private and commercial pilots. This means that the content and terminology are slanted toward aviators. It is nonetheless very useful to everyone in the affected area; it may just require a little "getting used to." The information given is comprehensive and up-to-date. A synopsis provides a summary of weather conditions in the area. Route forecasts give weather conditions within a 250-mile radius or along a 50-mile-wide band of selected flight routes (usually up to four). Observations of current weather conditions at nearby weather stations are also provided. In addition, pilot reports, weather-radar information, and severe weather warnings are included in the taped broadcasts that are continually repeated and updated every hour.

With the air weather broadcasts and the VHF weather radio so accessible, there can be little reason for being caught unawares.

Appendix D

A Backpack Weather Station

Having some simple and inexpensive weather instruments along can make almost any outing more interesting, more fun, and more safe. A few weather observations can easily be made without special instruments, such as cloud cover, cloud type, and wind direction. Others require the use of instruments if reliable and useful observations are needed. These include temperature, humidity, wind speed, and pressure and altitude. The following paragraphs describe instruments that meet several criteria appropriate to outdoorsmen: ruggedness, compactness, light weight, and reasonable cost. In many cases, other instrument types can be found that may perform equally well. Equipment can be ordered through science catalogs or purchased at outdoor equipment stores. A representative listing of manufacturers is offered on page 234.

Several types of temperature measurements are often desirable. In addition to the current air (or ground) temperature, it is often useful to know the high and low for the day. Special pocket thermometers are available for measuring the current temperature. They are ideally suited for outdoors because of the protection provided by the metal housing. They will not, however, provide daytime highs or nighttime lows without being continually watched. For this, you need what is called a *max-min thermometer*. Several types are available. The one illustrated on page 233 moves spring-loaded metal indices to the levels of the highest and lowest temperatures incurred during *any* time period. A small magnet is then used to reset the indices.

Humidity is easily measured with a *sling psychrometer,* consisting of a wet-bulb thermometer and dry-bulb thermometer mounted on a metal plate. Attached to the plate via a flexible coupling is a grip or handle. You first wet the wick (with distilled water) that covers the bulb of the wet-bulb thermometer, and then twirl the sling until the wet-bulb temperature stabilizes. A convenient pocket table converts the wet- and dry-bulb temperatures to the corresponding values of relative humidity and dew point. The pocket sling psychrometer illustrated here is only about 7 inches long and is equipped with a carrying sheath for protection. With a sling psychrometer along, it is not necessary also to have a simple thermometer.

Temperature and humidity information is far more useful if the altitude and atmospheric pressure are also known. A small aneroid altimeter/barometer is available that weighs only 3 ounces and is about $2\frac{1}{2}$ inches square and less than 1 inch thick. It comes with a protective leather case and a neck strap. Various models are available; the basic one measures to altitudes of 15,000 feet, while others span even greater altitudes.

The four selection criteria are not easily met in the case of a field-portable wind meter. Size, weight, and ruggedness are more easily satisfied than reasonable cost. There is a quantum leap in cost between the simple wind meter shown on page 233 and the more elaborate, compact anemometers that cost hundreds of dollars. But for the rough estimates that may be required, the Pitot tube-type sensor is adequate. The small size, light weight, and low price make it a wise choice. Yachtsmen who depend on accurate wind measurements in racing situations should, however, abandon the simple wind meter for the more precise cup- or propeller-type an-emometers illustrated in Chapter 1.

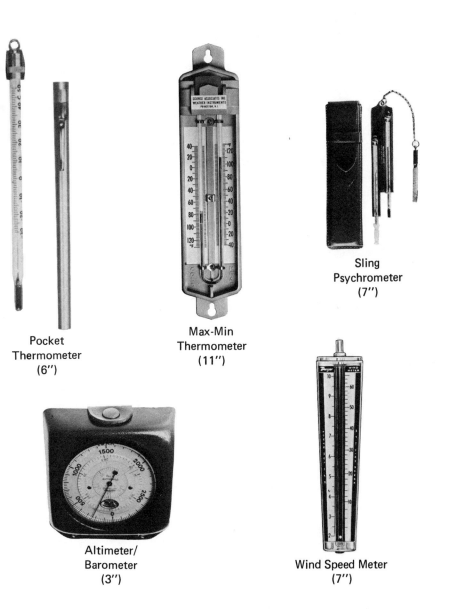

Pocket
Thermometer
(6")

Max-Min
Thermometer
(11")

Sling
Psychrometer
(7")

Altimeter/
Barometer
(3")

Wind Speed Meter
(7")

Backpack weather station. Courtesy of Dwyer Instruments (wind meter) and Science Associates of Princeton, N.J. (thermometers and altimeter/barometer)

Some Sources of Portable Weather Instruments

Edmund Scientific Company
101 East Gloucester Pike
Barrington, NJ 08007

Airguide Instrument Company
2210 Wabansia Avenue
Chicago, IL 60647

Dwyer Instruments
P. O. Box 373
Michigan City, IN 46360

L. L. Bean, Inc.
1042 Casco Street
Freeport, ME 04033

REI Co-op
P. O. Box C-88127
Seattle, WA 98188

Belfort Instrument Company
1600 S. Clinton Street
Baltimore, MD 21224

Science Associates, Inc.
230 Nassau Street
Princeton, NJ 08540

Weather Measure Corporation
P. O. Box 41257
Sacramento, CA 95841

Sierra Weather Instruments
   Corporation
P. O. Box 771
Nevada City, CA 95959

As mentioned earlier, outdoor equipment stores as well as nautical and aeronautical suppliers would be likely sources of portable weather instruments also.

References and Suggested Reading

BATTAN, LOUIS J. *The Nature of Violent Storms.* Science Study Series. Garden City, N.Y.: Doubleday and Company, Inc., 1961.

BROOKS, C. E. P. *Climate in Everyday Life.* London: Ernest Benn Limited, 1950.

DABBERDT, WALTER F. *The Whole Air Weather Guide.* Los Altos, Calif.: Solstice Publications, 1978.

DONN, WILLIAM L. *Meteorology.* New York: McGraw-Hill Book Company, 1965.

FEAR, GENE H. *Surviving the Unexpected Wilderness Emergency.* Tacoma, Wash.: Survival Education Association, 1975.

GEIGER, RUDOLF. *The Climate Near the Ground.* Cambridge, Mass.: Harvard University Press, 1965.

HALPINE, C. G., and TAYLOR, H. H. *A Mariner's Meteorology.* Princeton, N.J.: D. Van Nostrand Company, Inc., 1956.

IPPEN, ARTHUR T., ed. *Estuary and Coastline Hydrodynamics.* New York: McGraw-Hill Book Company, 1966.

JOHNSON, N. K., and DAVIES, E. L. "Some measurements of temperature near the surface in various kinds of soils," *Quarterly Journal of the Royal Meteorological Society,* vol. 53, 1927.

KUETTNER, J. P. et al. *Handbook of Meteorological Forecasting for Soaring Flight.* Technical Note 158, WMO No. 495. Geneva, Switzerland: World Meteorological Organization, 1978.

LANDSBERG, HELMUT. *Physical Climatology.* DuBois, Pa.: Gray Printing Co., Inc., 1962.

PERLA, RONALD I., and MARTINELLI, M., JR. *Avalanche Handbook.* U.S. Department of Agriculture, Forest Service, Agriculture Handbook 489. Washington, D.C.: U.S. Government Printing Office, 1976.

RIEHL, HERBERT. *Introduction to the Atmosphere.* New York: McGraw-Hill Book Company, 1965.

RUFFNER, JAMES A., and BAIR, FRANK E. *The Weather Almanac.* Detroit, Mich.: Gale Research Company, 1977.

SCHROEDER, MARK J., and BUCK, CHARLES C. *Fire Weather.* U.S. Department of Agriculture, Forest Service, Agriculture Handbook 360. Washington, D.C.: U.S. Government Printing Office, 1970.

TREWARTHA, GLENN T. *An Introduction to Climate.* New York: McGraw-Hill Book Company, 1968.

WATTS, ALAN. *Wind and Sailing Boats.* Devon, England: David & Charles Limited, 1973.

WATTS, ALAN. *Basic Windcraft—Using the Wind for Sailing.* Devon, England: David & Charles Limited, 1976.

Index

Illustrations are noted in *italics*.

Absolute. *See* temperature measurement
advection, 5, 9–10, 18, 43, 45, 51, 124
aerosols, 31
air. *See also* humidity; visibility
    density, 123, 154
    and heat conduction, 5–6, 9–11, 15, 18
    saturation, 20–21, 112, 197–98
    stability, 119–20, 192, 207
albedo, 40
altimeter, 235, 236
altitude. *See also* precipitation; pressure;
    temperature; wind speed
    measurement, 234–36
    and radiation, 153–54
altocumulus, 57, 60, 91
altostratus, xi, 57, 59–60, 85
anemometer, 16, *17*, 133, 135–36, 235
anticyclone, 63, 64. *See also* high-pressure
    systems
arctic smoke. *See* fog, steam
atmospheric pressure. *See* pressure
aurora
    australis, 36, *39*
    borealis, 36
    corona, *39*
    multiple-band, *39*
    rayed arc, *39*

barometer, 11–12, 80, 138, 235–36
barometric pressure, 80
Beaufort scale, 19, 122, 134
boat wind, 134, 136
boiling point, 5
bora, 179, 180
breezes. *See also* sea breeze front
    over lakes, 121, 124
    over land, 121, 123, *125,* 127
    in the mountains, 164, 168
    over sea, 51, 121, 123–24, 125

calorie, 4, 47–48
canyons, 152, 163
Celsius. *See* temperature measurement
Centigrade. *See* temperature measurement
chinook, 177–79
cirrocumulus, 56, 60
cirrostratus, xi, 56–57, 60, 85
cirrus, xi, 56–57, 60, 73
clouds, 100, 124, 127, 153, 161, 163, 190,
    209.
    See also names of individual cloud
      types; funnel clouds
    cirriform, 55
    cumuliform, 55, 61
    and energy transfer processes, 51
    and frontal systems, 84, 85, 91
    identification of, 55, 60–61
    and low-pressure systems, 64, 81, 84
    as precipitation indicators, 54–55, 57,
      59–61, 73, 177
    stratiform, 55, 91
    and temperature, 5, 22, 54
    and tornadoes, 103
    as wind speed and direction indicators, 14,
      54
cold fronts. *See* fronts, cold
condensation, 21, 40, 45–46, 66–67, 88, 90,
    161, 177–78
condensation nuclei, 112
conduction, 6–9, 18, 40–48, 50–51, 158–60,
    224–26, 228
convection, 44–45, 54, 58, 156, 188, 224–25,
    228
Coriolis effect, 13–14, 126
corpuscular radiation, 36
crests, of waves, 128, 130
cumulonimbus, 26, 58–59, 61, 84–85, 87–88,
    93, 137, 195, 205
cumulus, 16, 58, 59, 61, 73, 81, 84–85, 87,
    93, 95, 119, 126–27, 137, 153–54,
    163–65